Anatomy

A Matter of *Death* and *Life*

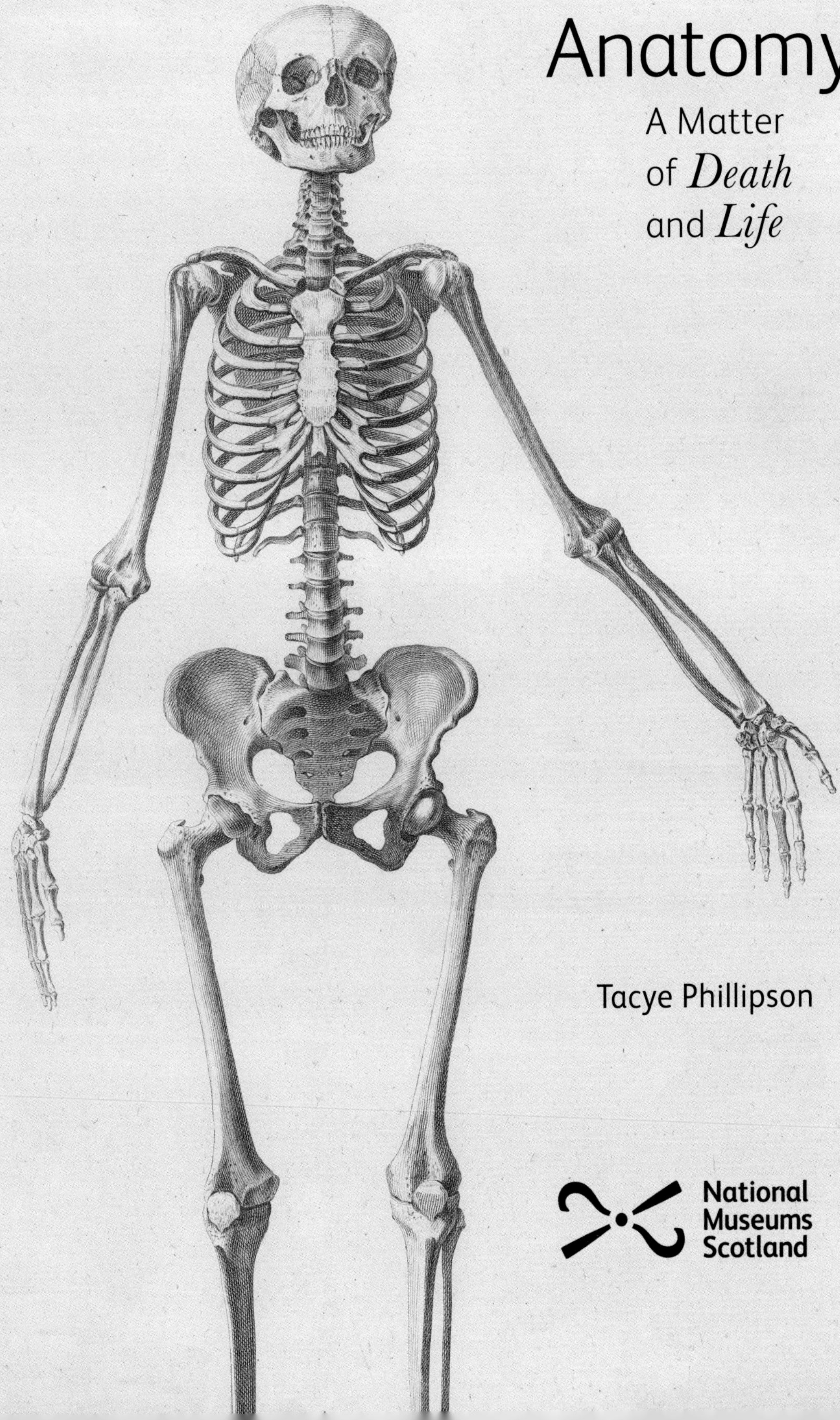

Tacye Phillipson

National Museums Scotland

Anatomy
A Matter of *Death* and *Life*

Exhibition at
National Museum of Scotland
Chambers Street
Edinburgh EH1 1JF

www.nms.ac.uk

2 July–30 October 2022

Exhibition kindly sponsored by
Baillie Gifford Investment Managers

Book published in 2022 by
NMS Enterprises Limited – Publishing
a division of NMS Enterprises Limited
National Museums Scotland
Chambers Street
Edinburgh EH1 1JF

www.nms.ac.uk

Text, photographs and images
© National Museums Scotland 2022
(unless otherwise acknowledged)

No part of this publication may be reproduced, stored in a retrieval system, or transmitted in any form or by any means, electronic, mechanical, photocopying, recording or otherwise, without the prior written permission of the publisher.

The right of the author to be identified as the owner of this work has been asserted by them in accordance with the Copyright, Designs and Patents Act 1988.

British Library Cataloguing in Publication Data
A catalogue record of this book is available from the British Library.

ISBN 978 1 910682 46 3

Typesetting by NMS Enterprises Limited – Publishing.

Cover design by Mark Blackadder, based on design by NMS Exhibitions and Design department.
Photography (unless otherwise credited) by Amy Campbell, Neil McLean and Olga Tjukova of NMS Photography.
Printed and bound in Great Britain by Bell & Bain Limited, Glasgow.
Cover: Adapted from a lithograph of the circulatory system by J. Maclise, 1841/1844. Image: Wellcome Collection. Attribution 4.0 International (CC-BY-4.0).
Page 1: Taken from a treatise on osteology by Alexander Monro *primus* (1726). Image: Wellcome Collection. Attribution 4.0 International (CC-BY-4.0).

This product is made of material from well-managed forests and other controlled sources.

The exhibition, and this book, contain many anatomical objects, images and models from a broad time period. The people whose dissected dead bodies were the models for these objects and images are unlikely to have given their consent.

For a full listing of
NMS Enterprises Limited – Publishing
titles, and related merchandise, go to:

www.nms.ac.uk/books

Anatomy
A Matter of *Death* and *Life*

Introduction 4

History of Anatomy 7

Edinburgh and Anatomy 33

The West Port murders 61

An end to grave robbing 92

Further reading 94

Acknowledgements 95

Introduction

In late 1828 and into the beginning of 1829, a news story originated in Edinburgh which horrified and outraged all who heard it. People, at first in numbers unknown, had been murdered in and around Edinburgh's West Port district and their bodies bought by a respected anatomist, Dr Robert Knox, as subjects for dissection. Referred to at the time as the West Port murders – more commonly known now as the Burke and Hare murders – these events remain notorious to this day, and the idea of being killed purely for your body remains a chilling one. But was there any explanation behind these deaths? Why did these murders take place in Edinburgh? And were the events unique?

Our understanding of anatomy is based extensively on studies of dead people: the dead helping the living to learn more about themselves. Knowledge of anatomy was, and is, considered a key part of medical education, enabling better diagnosis and treatment. However, anatomical knowledge and surgical ability to make use of this in practice have not advanced at the same pace. Leonardo da Vinci studied heart valves more than four centuries before the first successful operation took place in 1925. While this could not have happened without an excellent anatomical understanding, for centuries such knowledge was of no practical use for surgical treatment. This does not mean it was medically useless, however. Even without the ability to repair anatomical defects, knowledge about them could be useful: for instance, in identifying the cause of death through post-mortem investigation, providing answers increasingly sought by the relatives of deceased individuals. Comparison of post-mortem anatomical findings with observed symptoms expanded the diagnostic ability of doctors and provided information for prognosis and non-surgical treatment. The Scottish surgeon and anatomist John Hunter (pages 50–53) confidently diagnosed himself with heart failure, in which he was an expert, and acknowledged, 'My life is in the hands of any rascal who chooses to annoy and tease me'. Sadly he was unable to prevent his death at 65 years of age in 1793 – the autopsy confirmed his self-diagnosis.

Opposite: Papier-mâché anatomical model by Auzoux, France, 1879

From the 1820s, the Auzoux workshops supplied papier-mâché anatomical models to medical schools worldwide. These models could be taken apart or 'dissected' repeatedly. This example was bought by the University of Aberdeen in 1879 and separates into 92 pieces. The models were popular and practical teaching aids, but they could not replace the knowledge or surgical experience gained from dissecting real bodies.

© University of Aberdeen, Scotland

Anatomy is closely allied to physiology: what parts of the body do and how each part works. While the structures are present for any dissector to reveal, it is the understanding of what is being sought – for guidance, confirmation of existing knowledge, or following a hypothesis or query – that helps to separate and make sense of the closely interlinked structures of the body. This, as well as the physical differences between people, is why anatomists had to carry out multiple dissections on many different bodies.

There are two main closely-related justifications for anatomical dissection. The first is research: to acquire understanding about the human body. The second is teaching: to impart existing knowledge either through demonstration, or – in increasing numbers as practices changed – through students carrying out dissection themselves.

For centuries, anatomists relied on the bodies of deceased people from poor backgrounds, taken without consent, for most of their dissections. Drawings, models, preserved specimens and – more recently – medical scans, all played a part in anatomical education and reduced the use of bodies. However, these could not entirely replace the knowledge gained from the dissection of real bodies, nor the surgical experience acquired by carrying out dissections. Today, anatomists in the United Kingdom dissect only donated bodies, given with consent.

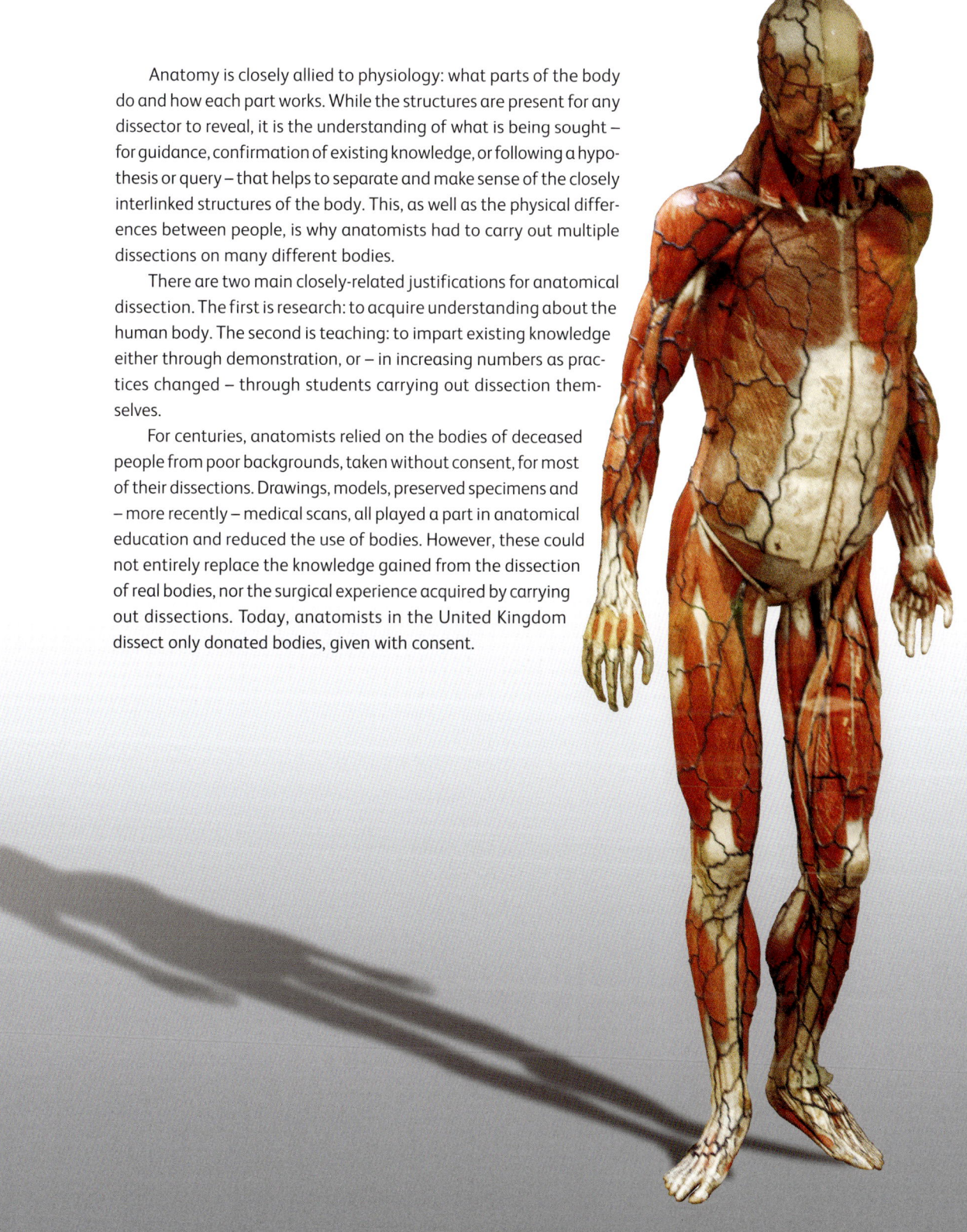

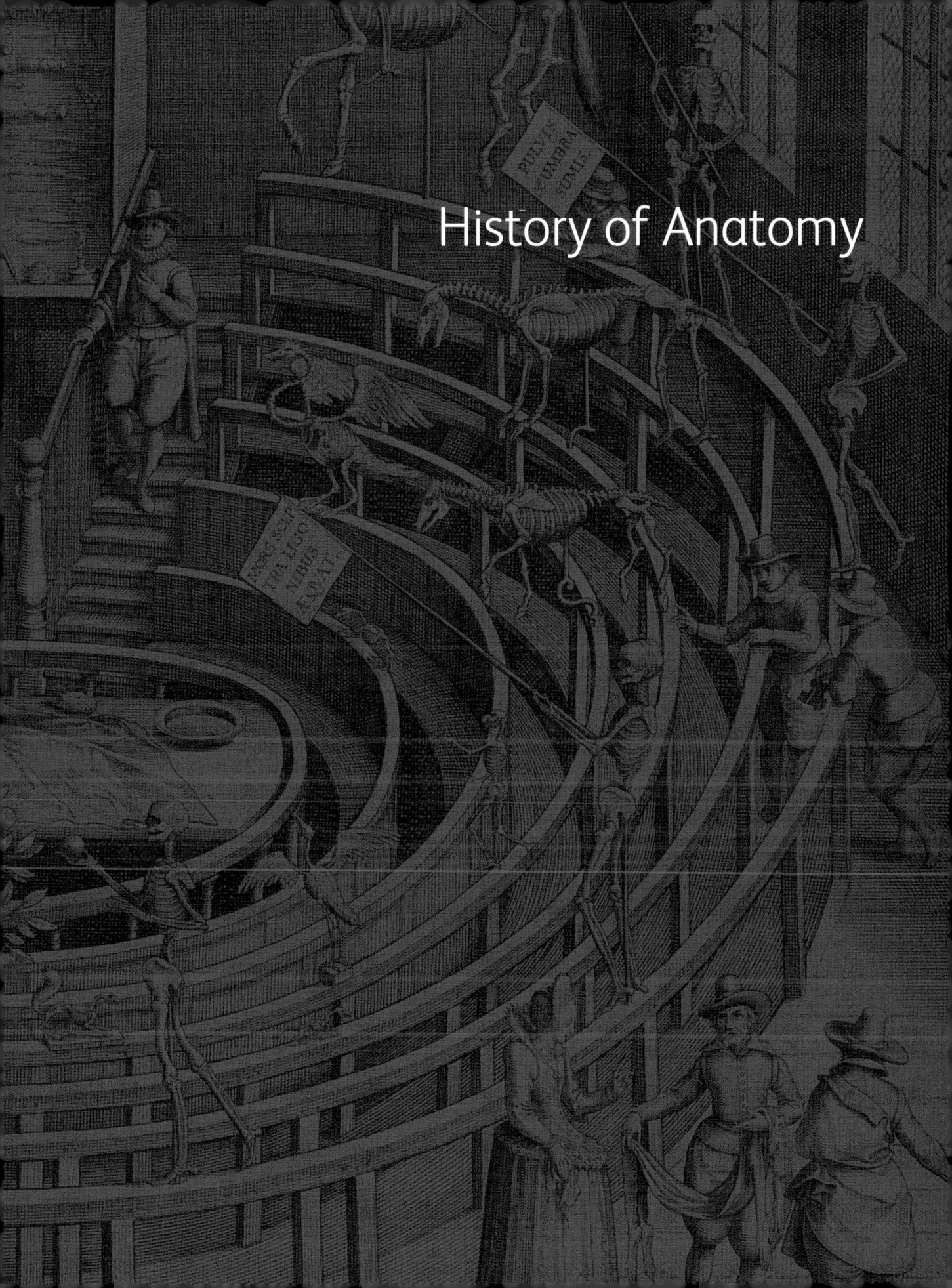

History of Anatomy

There is not a single point at which it can be stated that the study of anatomy began. Perhaps an individual acting as a healer, midwife, warrior, sculptor or embalmer would have found a practical understanding of human anatomy useful. Traces of this knowledge – such as bones that have been set and healed – appear in the archaeological record pre-dating any written anatomical texts. Then, as written and pictorial records of human achievement and knowledge increased, so too did the records that relate to anatomy.

In Medieval and Renaissance Europe, education – including anatomy – was heavily based on the reading and interpretation of works by classical scholars. In the case of anatomy, this was frequently from Galen (Claudius Galenus), a Greek physician of the Roman Empire in the second century CE, whose writings were prolific due to the employment of scribes to take down his words as he talked to students or peers. He made anatomical observations and carried out demonstrations based mainly on the dissection of animals and the external examination of dead people, rather than the dissection of human bodies. In his time, even slaves and slaughtered gladiators appear to have had social protection from being dissected, although no legal basis for this has been identified.[1]

In some areas of Renaissance Europe, limited dissection of human bodies was accepted as part of anatomical teaching from the mid-thirteenth century. The Catholic Church considered it to be a desecration of the body, but authorised the practice in certain cases of executed criminals, on condition the body was buried afterwards. A typical anatomy lesson would involve the teacher reading from, and expounding upon, a book – often one of Galen's works in Latin translation (from the Greek or Arabic versions) – while a surgical assistant carried out the menial and messy dissection and demonstration. Such was the strength of Galen's writings as the foundation of anatomical and physiological education that observations not matching his descriptions were frequently passed over or explained away.

Anatomy flourished most where the economic climate allowed energy to be devoted to all kinds of learning and culture. The social, religious and political elite became more accepting of the dissection of bodies and permitted anatomists their studies, either for the argued medical benefits that anatomically-trained physicians and surgeons would bring, or for the advancement of knowledge for its own sake and the better understanding of man, God's creation.

Science and art of Anatomy

In the early sixteenth century, scholars were developing new ways of thought and what we now call science. They emphasised the value of observation – seeing for yourself – instead of relying on the preserved knowledge of Greeks and Romans recorded in classical texts. Dissection was central to this approach as it applied to knowledge of the human body, and the Latin phrase *nosce te ipsum*, 'know thyself', became associated with anatomical study, linking the anatomist to the dead body on the dissection table.

Surviving records reveal that the study of anatomy in Europe was far from consistent, with a mix of individual enterprise and limited organised teaching. Obviously, present knowledge of how anatomical study took place is heavily influenced by available accounts, but there are also hints of unrecorded dissections. In 1505, the Incorporation of Barber Surgeons of Edinburgh, forerunner to the Royal College of Surgeons of Edinburgh, was founded. The Town Council granted the Barber Surgeons the body of 'one condemned man' a year for dissection, in the teaching of their apprentices. Unfortunately there is no information about whether these bodies were claimed, or indeed anything about dissections. However, that a request was made in the first place strongly indicates they were already happening in Edinburgh and this was not the introduction of a novel educational practice. Such dissections were probably taking place privately in a surgeon's home, with the bodies acquired by whatever means.

Undocumented dissections would not have been unique to Edinburgh, but part of the medical landscape across Europe. Observing or carrying out dissections provided useful experience for medical practitioners, many of whom were unqualified. Indeed, in numerous cases no relevant qualifications had yet been introduced. In time, more educational and apprenticeship routes became formalised, leading to organised anatomical teaching and qualifications for professionals such as surgeons and midwives.

Qualifications in medicine at this time were awarded by a small number of universities. This was often better documented, at least in recording the names of those appointed to teach. In Europe there were only a few dozen universities, not all offering medical teaching; anatomical teaching and dissection could only take place openly where religious and civil authorities tolerated the practice.

Aside from medical training, anatomical study was used by artists to improve technical and observational skills. Others were motivated by a general interest in advancing their understanding of the human body, although they did not expect to put this to practical use. The polymath Leonardo da Vinci falls into both of these categories as his anatomical notebooks delve deeper into the structure and function of the body than would have been useful for his art.

Art and Anatomy

Anatomy was, and is, a part of medical education, but it is also of great interest to artists. There is mutual benefit between these fields. The close focus required to draw a subject accurately can enhance understanding, or highlight areas where there is a lack of knowledge. Similarly, an understanding of what you are seeing assists in the drawing of it. Artists studied the anatomy of bones, muscles, ligaments, fat and skin to inform their own work – to pose, draw or sculpt realistic (or idealised) human forms. In this way they could comprehend how the body functions and convey an extra sense of realism. For similar reasons, some artists were also interested in animal anatomy, especially of horses.

This relationship between anatomy and art was described by the sixteenth-century Italian painter and writer, Giorgio Vasari:

> *Again, having seen bodies dissected one knows how the bones lie, and the muscles and sinews, and all the order and conditions of anatomy, so that it is possible with greater security and more correctness to place the limbs and arrange the muscles of the body in the figures we draw.*[2]

Three-dimensional anatomical sculptures produced by artists were useful studies for flat art, statues and anatomists. These *écorché*, or flayed figures – showing figures without their skin and fat to reveal muscles and tendons more clearly – were sculpted in wax or plaster, sometimes duplicated by being cast in bronze.

Anatomists drew upon artists and their skills to record and share their own observations. Artists are still used today to produce illustrations that show particular anatomical features more clearly than unenhanced photographs.

Art is likewise important for the study of the history of anatomy. Depictions of dissections being carried out provide insight into the study and teaching of the latter, and of the places where dissections took place. However, the interpretation of such images requires a degree of care as they convey a mix of representation and elaboration.

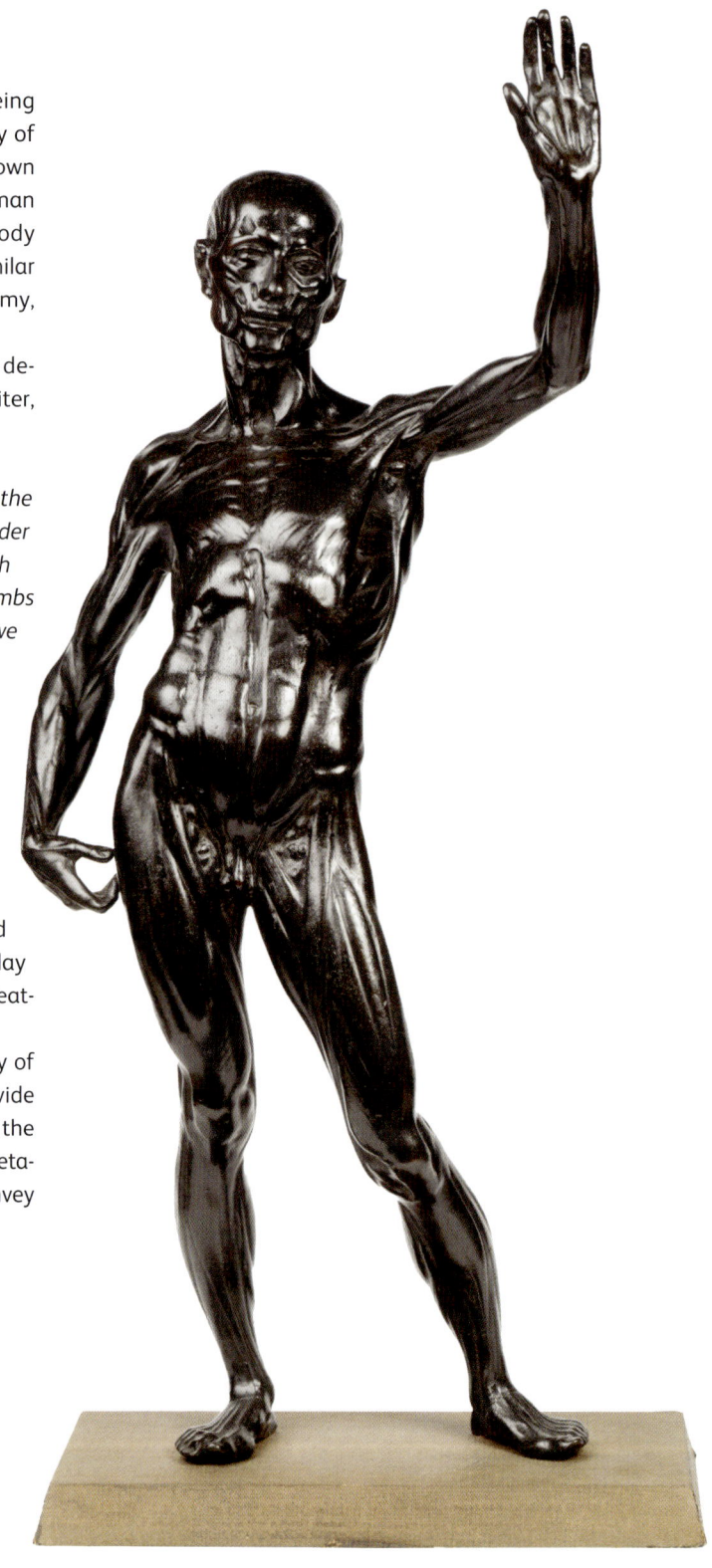

Écorché figure of a man, Italy, c.1580

Flayed, or *écorché*, figures clearly show the muscles and tendons without being obscured by layers of fat and skin. This sort of anatomy was particularly relevant to artists depicting people.

© Victoria and Albert Museum, London

Leonardo da Vinci

Leonardo da Vinci (1452–1519) is today acknowledged as a highly skilled and pioneering anatomist, in addition to many other achievements. This reputation rests on his anatomical drawings, which were of such artistic quality they were acquired for the British Royal Collection by 1690. Their anatomical excellence, however, was mostly overlooked until Dr William Hunter studied them in 1773. Publicising the drawings and the anatomical skills, Hunter stated that da Vinci had 'been overlooked, because he was of another profession, and because he published nothing upon the subject'.[3]

The drawings and annotations reveal skill and understanding well in advance of those found in published works of the time, sometimes even contradicting the established facts set out by Galen. They show both an artist's practical interest in visual anatomy and a mechanistic interest in physiology. However, as they remained unpublished in his private manuscripts, da Vinci's influence on the development of anatomy was not significant. He had planned to publish a book, but it never came to fruition.

Access to bodies for dissection was limited and variable throughout da Vinci's life. He somehow acquired a human skull in 1489, and was thought to have dissected over thirty human bodies. He also dissected animals, which were much more readily available. In his work he illustrated that the heart had four chambers, not two as previous anatomical texts stated, and his unpublished notes contained information about the flow of blood through the heart valves that was only rediscovered about five hundred years later.

At the time, and for centuries afterwards, the detailed working of heart valves was perhaps of little practical use to physicians or surgeons in the treating of patients, but da Vinci's concerns were not solely with helping the living medically, but in identifying the cause of death. In the winter of 1507/8 he visited a male patient in the Santa Maria Nuova hospital, Florence, who claimed to be over a hundred years old. Hours later, the man died and was later dissected by the artist. (It is thought that the man was aware of this eventuality, in exchange for the care received during his life.) Da Vinci wrote later, 'I made an anatomy of him to see the cause of a death so sweet, and found that it proceeded from weakness through the failure of blood and of the artery that feeds the heart and the other lower members, which I found to be very dry, shrunken and withered'.[4] This was the first dissection to document both the process of arteriosclerosis and cirrhosis of the liver. The anatomy of the centenarian's blood vessels was then compared to that of a young boy, documenting the differences between people that are key to understanding where anatomy is normal and where it is the source of symptoms or illness.

In 1509, Leonardo da Vinci met and collaborated with the Professor of Anatomy at Pavia University, Marc'Antonio Della Torre, who taught the subject using dissection. Della Torre's untimely death from plague two years later seems to have played a significant part in da Vinci setting aside anatomical studies and his work failing to reach publication. The professor himself is known to have produced anatomical works and to have been highly respected in his time, but none of his works survive. Perhaps gaps in our present knowledge of anatomical history may be down to discoveries that are simply unpublished or lost.

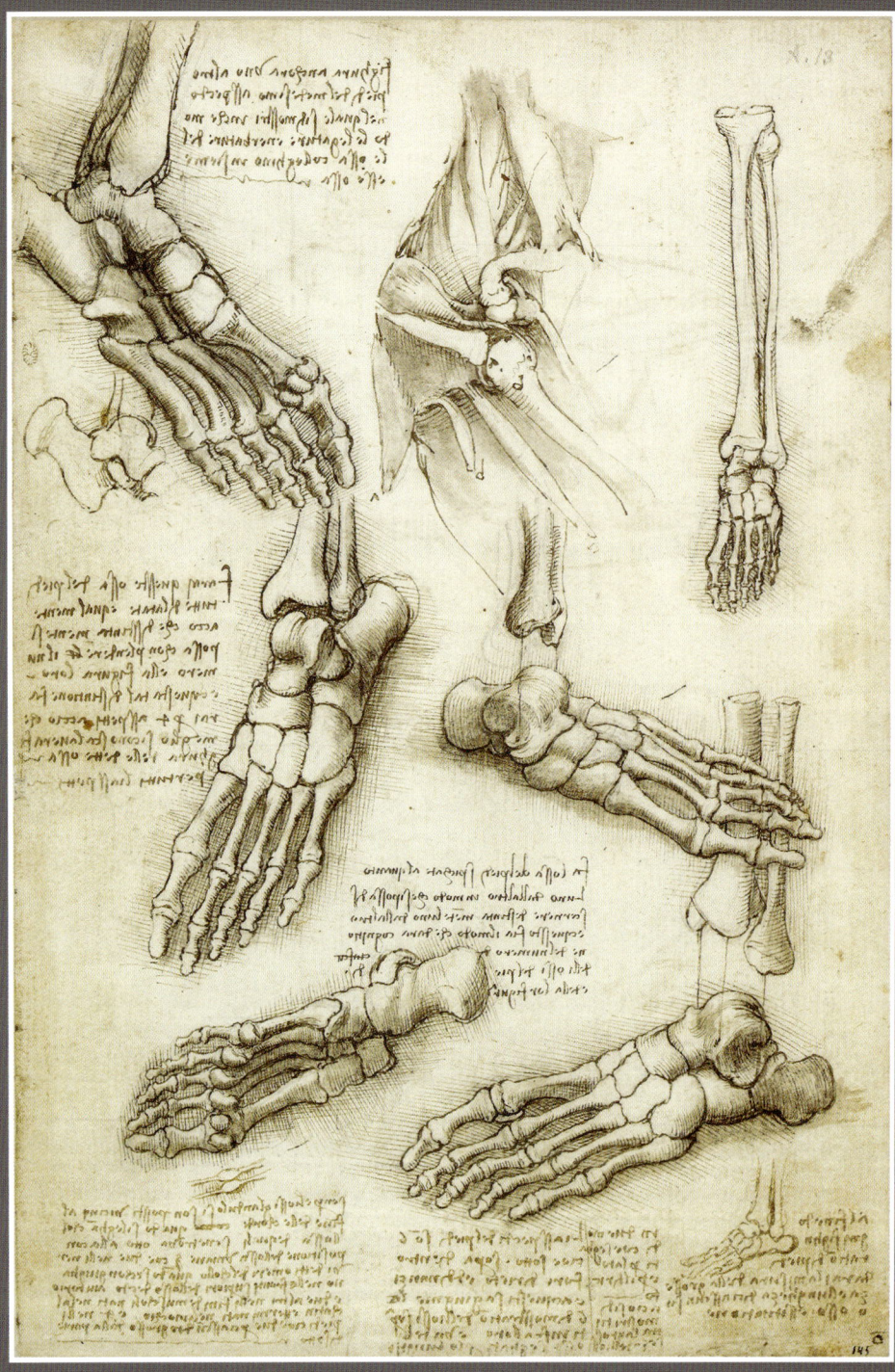

The bones of the foot and the shoulder, by Leonardo da Vinci, pen and ink with wash over traces of black chalk, c.1510–1511

Da Vinci captured every bone of the foot in this drawing, but the arch is incorrect. This may be because he was working from dried material.

Royal Collection Trust / © Her Majesty Queen Elizabeth II 2022

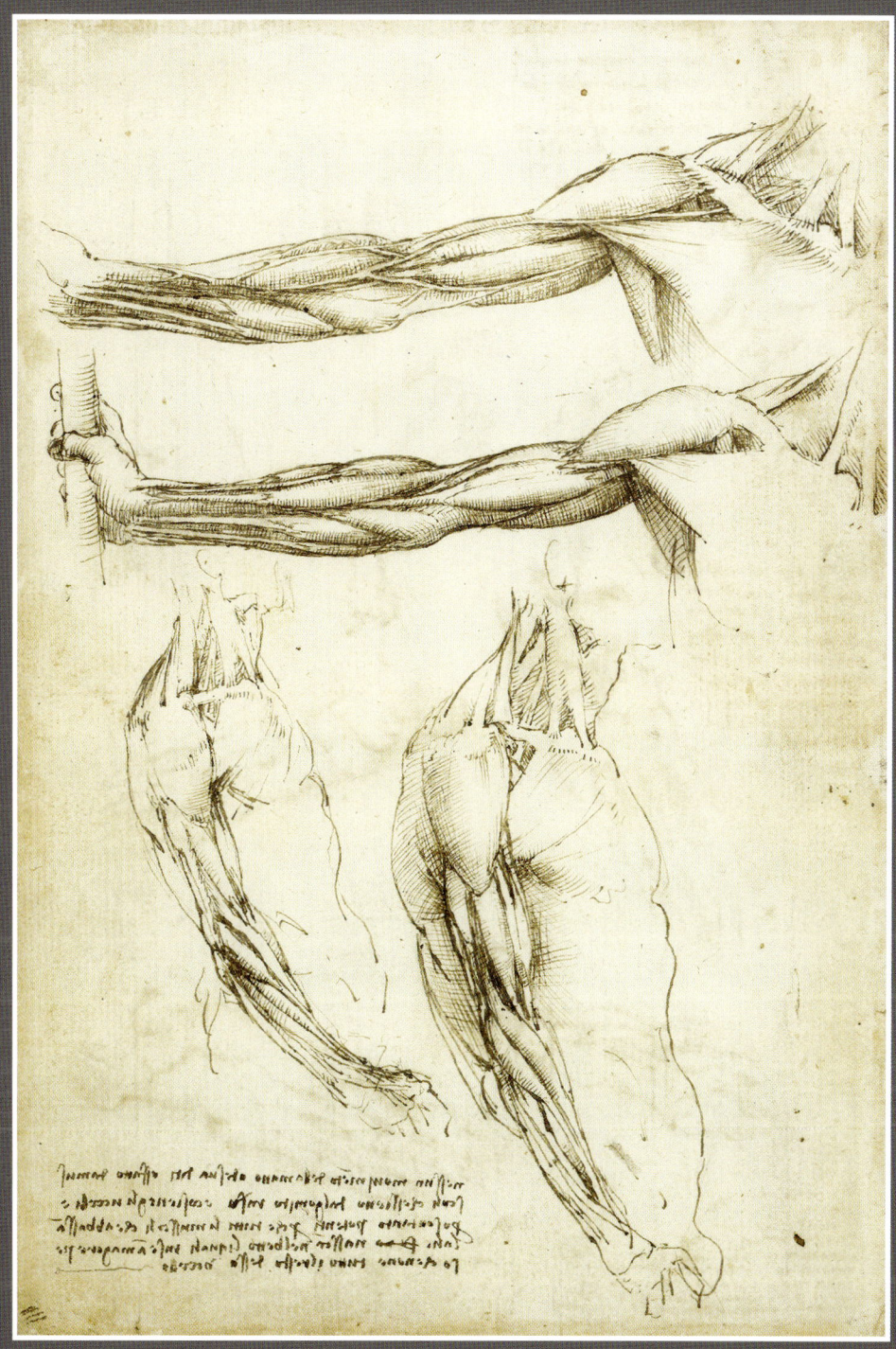

The veins and muscles of the arm, by Leonardo da Vinci, pen and ink over traces of black chalk, c.1510–1511

Da Vinci's notes in mirror writing here compare the muscles of a human arm to those in a bird's wing. This comparative anatomy helped him to understand better the way muscles function in the human arm.

Royal Collection Trust / © Her Majesty Queen Elizabeth II 2022

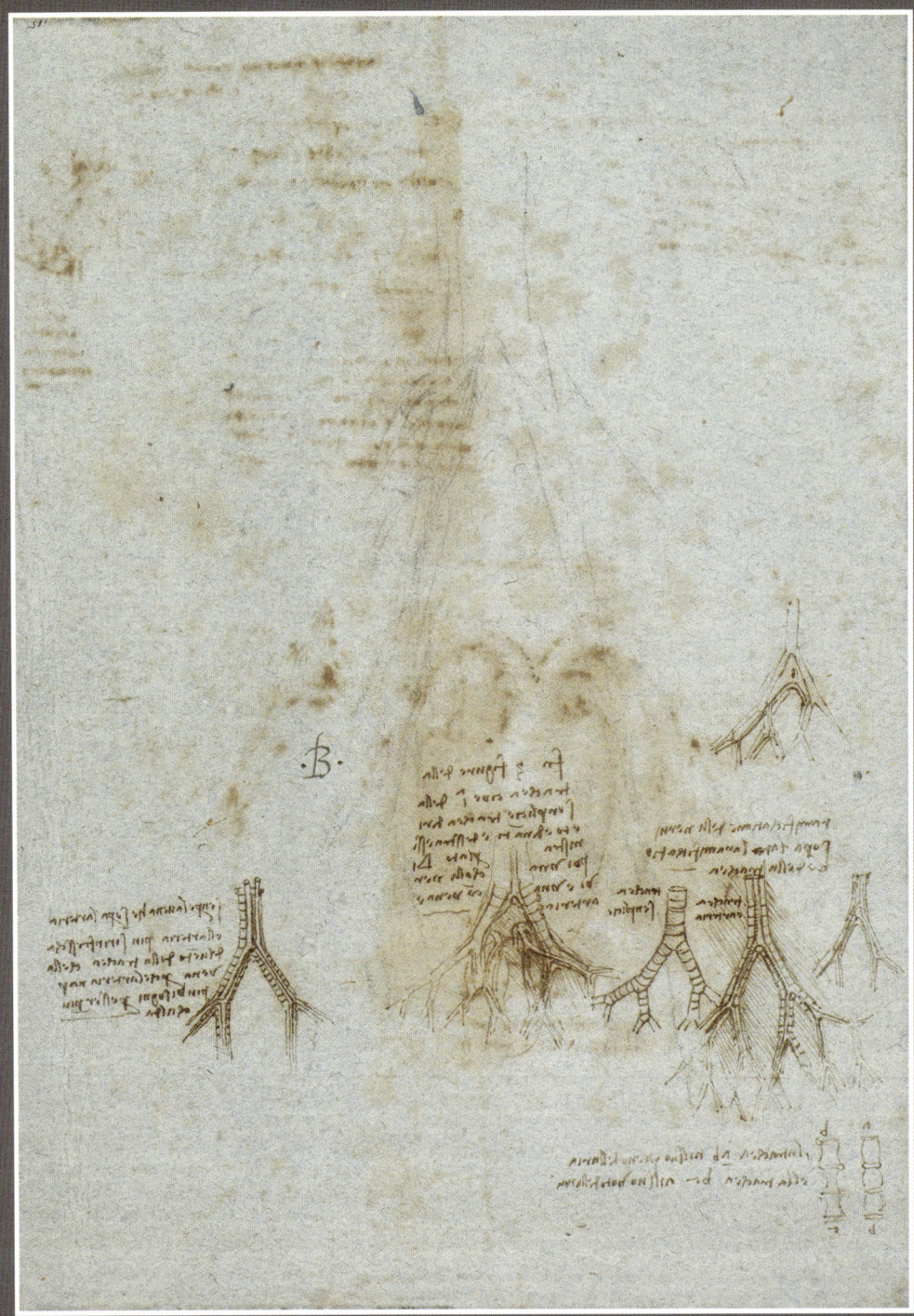

The trachea and associated vessels, by Leonardo da Vinci, pen and ink and black chalk, c.1511–1513

Multiple studies of the same part of the body – in this example the trachea (windpipe) – helped da Vinci to understand the differences in anatomy between different people. Paper at this time was made by using clothing rags, in this case giving the paper its blue colour.

Royal Collection Trust / © Her Majesty Queen Elizabeth II 2022

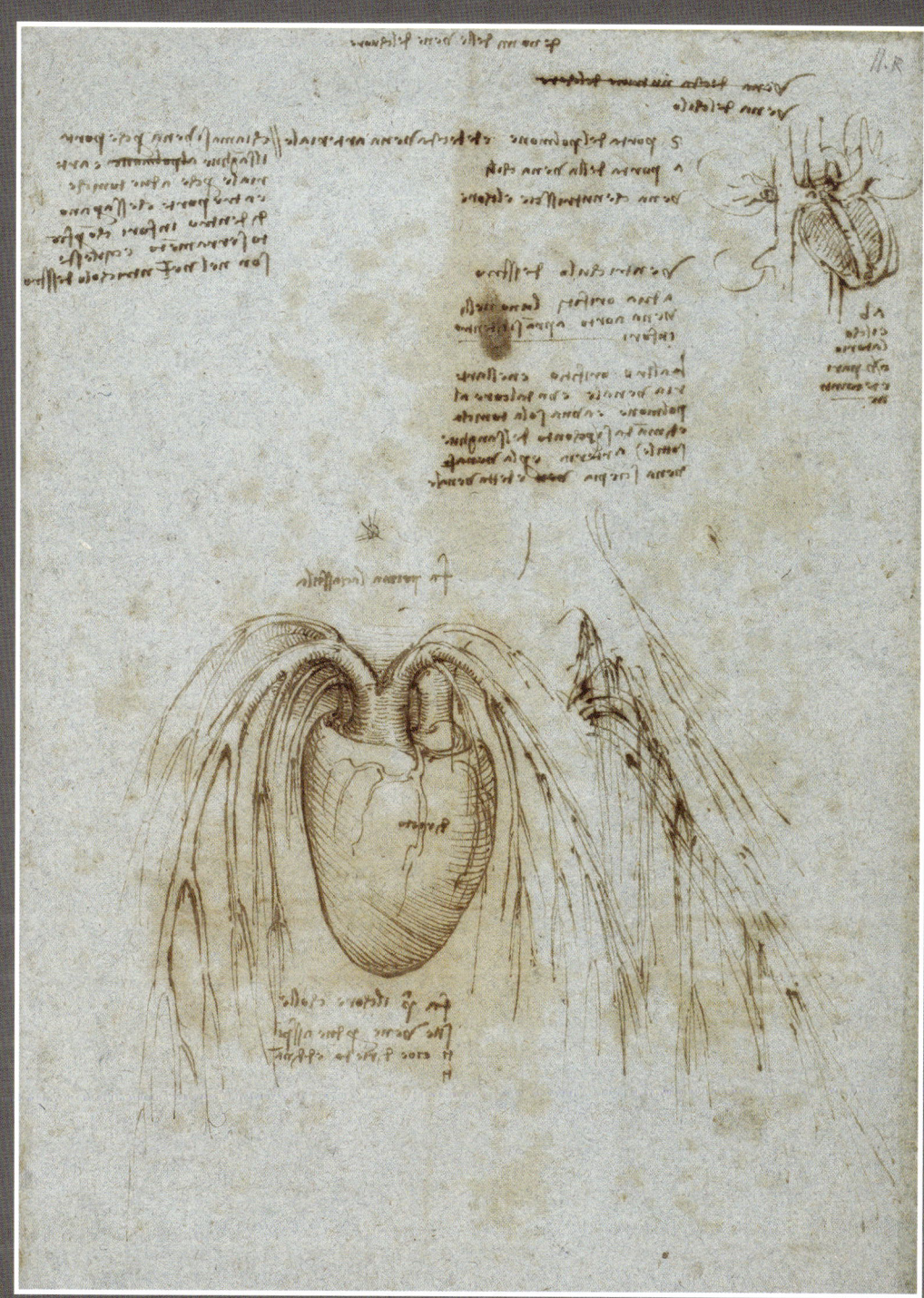

The heart and pulmonary vessels of an ox, by Leonardo da Vinci, pen and ink, *c.*1511–1513

Da Vinci could not always obtain human specimens. His notebook contains many dissections of other animals, in this case an ox, to try to understand the way the human body worked.

Royal Collection Trust / © Her Majesty Queen Elizabeth II 2022

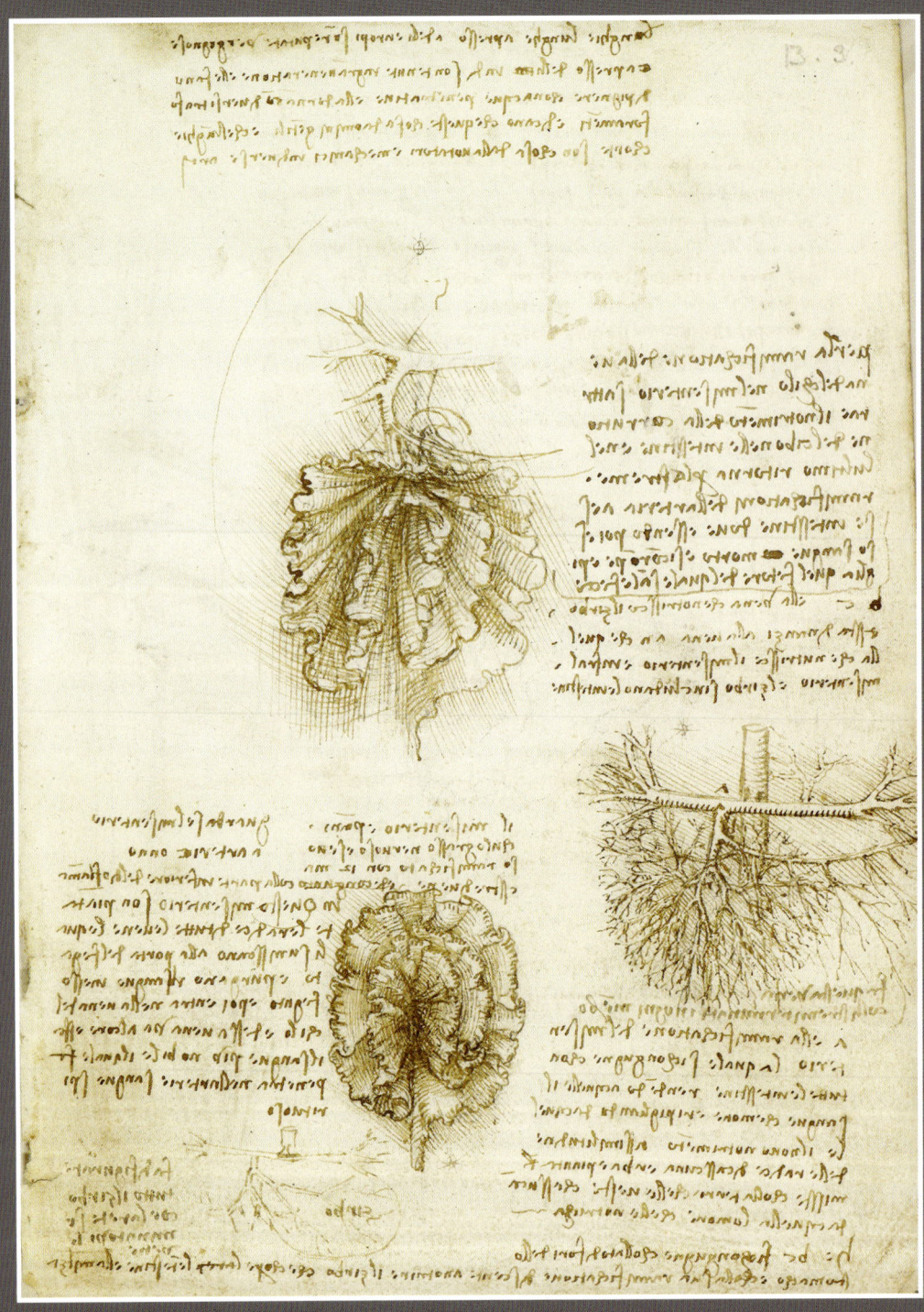

Dissection of the bowel, by Leonardo da Vinci, pen and ink over black chalk, *c*.1508

At this point in his career da Vinci had dissected at least ten bodies, comparing their anatomy. Immediately following the dissection of the centenarian, da Vinci dissected a two-year-old boy and 'found everything contrary to that of the old man'.

Royal Collection Trust / © Her Majesty Queen Elizabeth II 2022

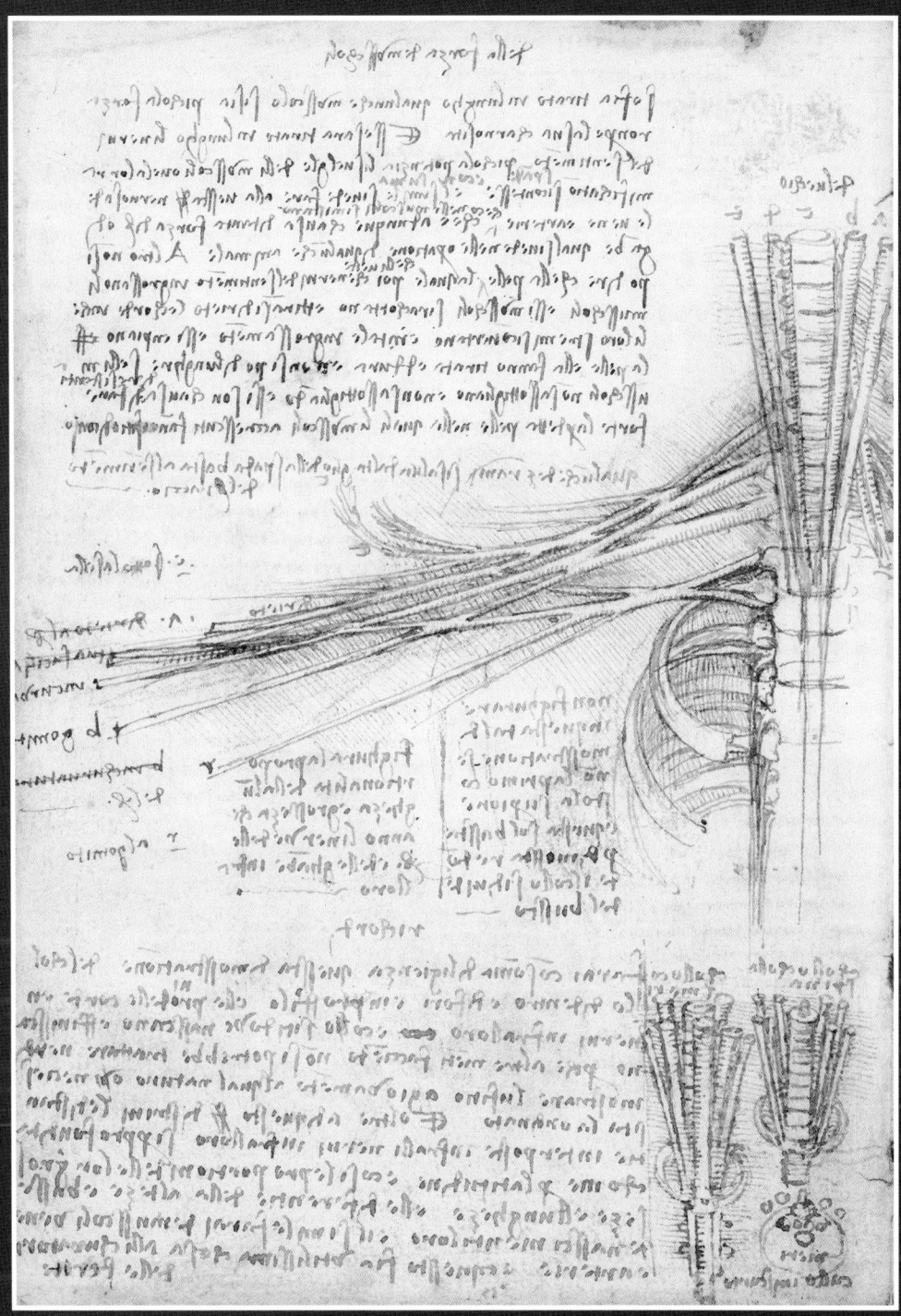

The brachial plexus, by Leonardo da Vinci, pen and ink over black chalk, *c.*1508

Da Vinci made several sketches of his dissection of the centenarian including this one of the brachial plexus, the group of nerves which travel from the spine to the arm. As part of this dissection, he also identified cirrhosis of the liver for the first time.

Royal Collection Trust / © Her Majesty Queen Elizabeth II 2022

Andreas Vesalius

De humani corporis fabrica [*On the Fabric of the Human Body*], written by Andreas Vesalius, was published in 1543. Vesalius was Professor of Anatomy at the University of Padua and his book is widely acclaimed as the foundation of modern anatomy. This is not only because of the anatomical observations contained within it, but the author's bold insistence upon giving priority to his own work over the established statements of Galen. Vesalius had the courage to take a controversial and public stance against some of his predecessor's observations, demonstrating where they were true of particular animals but different in humans. It is unlikely to be accidental that the frontispiece of the book shows Vesalius himself carrying out the public dissection of a woman's abdomen, while the surgical assistants who might previously have done this are relegated below the action. The book (in seven volumes) was large and expensive, and few students could have afforded to purchase a set for themselves. A cheaper abbreviated version was simultaneously published to increase the work's availability and dissemination. In this way, Vesalius hoped to ensure more sales of his own books over any unauthorised copies issued by publishers and anatomists.

While the frontispiece to the work is clearly not drawn from life, Vesalius describes the event it portrays. The body being dissected was that of a woman who was hanged for an unspecified crime (although he refers to her as a prostitute). She had attempted to delay her execution by claiming she was pregnant, but a midwife who examined her disagreed. After the execution, Vesalius was requested by the judicial authorities to confirm the midwife's observation.

The dissection would have taken place in a theatre constructed inside an existing room, attended by crowds of both the genuinely interested and those who were more attracted by the occasion than the learning. There is also a considerable degree of artistic licence and symbolism in the image. The dog and monkey, often themselves subjects for dissection, were unlikely to be running loose in the crowd, and while the naked man climbing round a column may have been a model for a lecture on surface anatomy, as observed in a living person, his positioning is less than plausible.

Vesalius' book was the finest illustrated work of anatomy published up to then, containing over two hundred woodblock illustrations. It established the concept that image, not just description, was key to communicating anatomical knowledge. The illustrations were drawn by an excellent artist, or artists, working under his direction. The senior contributor is thought to have been Jan van Calcar who trained with Titian, and who had produced the drawings for Vesalius' previous publication, *Tabulae anatomicae sex* (1538) [*with six anatomical plates*]. However, while van Calcar was credited in that first publication, relations seem to have cooled between Vesalius and himself and he is not named in *De humani corporis fabrica*.

Opposite: *De humani corporis fabrica libri septem*, by Andreas Vesalius, 1543, frontispiece

This frontispiece shows a large audience attending as Vesalius dissects an executed woman to verify she was not pregnant. She had said she was pregnant, which would have postponed her execution, but the midwife disagreed.

British Library, London, UK © British Library Board. All rights reserved / Bridgeman Images.

Vesalius complained about the difficulties of working with artists. Making anatomical drawings from dissected and decaying bodies was, unsurprisingly, not a popular commission. Nonetheless, the quality and accuracy of the illustrations are particularly outstanding where they show bones, muscles and tendons – the large physical anatomy of familiar relevance to an artist drawing the human form. The plates dealing with the nervous system, however, are considered lower quality – possibly reflecting the difficulty of transmitting Vesalius' understanding to the artist, rather than a lesser quality of dissection and preparation of the body.

The poses that were selected were carefully thought out for scientific rather than merely artistic reasons. They illustrate how the various parts are interconnected far more clearly than the stiff front and back views of a prone body. Vesalius used ropes and pulleys to position his subjects as he wished. This practice would have enabled the artist(s) to draw from the positions shown. As a consequence, the quality and recognised importance of the woodblocks was such that they were preserved until destroyed by bombing during the Second World War.

Vesalius was able to dissect the bodies of people who had been sentenced to death, but also wrote about taking the acquisition of bodies into his own hands. As a student in Paris and Louvain, fascinated with anatomy but dissatisfied with the resources available, he said that graveyards were an easy source of material, particularly the bones disturbed when digging new graves and relocated to a charnel house. He wrote openly – for the elite who were literate and wealthy enough to access his writings – about where bodies might be found if not provided by a co-operative judge. In addition to the anatomical illustrations, the illuminated capitals in his book depict men or *putti*, naked children, engaged in many of the activities of an anatomist: robbing graves, boiling bones, dissecting animals and humans, or hoisting them into position with ropes, but also setting broken bones and assisting in childbirth. Despite the implausibility of *putti* being featured in these scenes, they provide useful information about the practicalities of anatomical study.

Vesalius was only Professor of Anatomy from 1537–1542. Most of his influence in anatomy came from his work during that short period, which culminated in a move to Basel to supervise the printing of *De humani corporis fabrica*. This was followed by his appointment as court physician to the Hapsburg Emperor, Charles V – a highly prestigious post, but one where he had to get used to treating live people rather than dissecting the dead. His book was probably key to this appointment, and certainly significant to his reputation from then until the present day.

Above: *De humani corporis fabrica libri septem*, by Andreas Vesalius, 1543

The illuminated letters show *putti* engaged in the actions of an anatomist or physician. In 'Q' the *putti* are vivisecting a pig; while in 'T' they are suspending an animal, probably a dog.

Q. The Picture Art Collection / Alamy Stock Photo
T. AF Fotografie / Alamy Stock Photo

Opposite: *An anatomical dissection by Pieter Pauw in the Leiden anatomy theatre*, engraving by Andries Stock after a drawing by Jacques de Gheyn II, 1615

Pieter Pauw was the first Professor of Anatomy at Leiden, and responsible for the construction of its anatomical theatre. The audience are not entirely focussed on the dissection, and the drawing shows the influence of Vesalius' frontispiece including the skeleton, and dogs in the foreground.

Wellcome Collection. Attribution 4.0 International (CC-BY-4.0).

Anatomy theatres

Vesalius did not have a purpose-built anatomical theatre in which to teach, but the world's first was inaugurated at the University of Padua in 1595, decades after his tenure there. This theatre still survives. It was closely followed by one in Leiden in 1596 at the behest of Pieter Pauw, the University's first Professor of Anatomy, who had studied the subject in Padua and elsewhere. Pauw had forty prints based on Vesalius' work hung on the walls of the new theatre as illustrations for his lectures and for students to study without requiring textbooks, which were not yet widespread. The anatomy theatre was used for both human and animal dissection and also housed preserved specimens, especially skeletons. Not every anatomical lesson required a dissection.

Contemporary illustrations exist of the Leiden anatomy theatre. One from 1615 shows Pauw dissecting in a theatre probably shrunk by artistic licence. The skeleton, likely present in reality, has a banner reminding viewers of their mortality: *Mors ultima linea rerum* – death, the final boundary of things. Anatomy was acknowledging its close connection to the dead. Mortality would have been a familiar theme in the seventeenth century as life expectancy was much shorter than today.

Another image of the Leiden anatomy theatre, drawn c.1610 by Jan Cornelis Woudanus, exists in a few versions – one with a large audience and a dissection taking place; another without a lecturer, but with various visitors walking through the space. All have skeletons mounted around the theatre, including a prominent pair, Adam and Eve, in the foreground, and a serpent in a tree. This image probably combines a representation of the space with allegorical artistic embellishments.

Even when human or animal dissections were not taking place, the theatre attracted the curious. A catalogue of specimens kept in the Leiden theatre was published in English, Dutch, Latin and French, presumably for tourists as students were taught in Latin. Titled *A catalogue of all the cheifest rarities in the publick theatre and anatomie-hall of the university of Leyden* (1704), it includes the 'sceleton of an asse upon which sits a woman that killed her daughter' and the 'sceleton of a man, sitting upon an ox, executed for stealing of cattle', confirming there were indeed posed skeletons around the hall.[5] The publication, though later than the engravings, does not match exactly, and the skeletons were likely not posed as shown in the engravings, especially as they are not the same in the different versions of the image.

Anatomy theatre at the University of Leiden, 1610

Anatomy theatres were also used for animal dissections and were places to see and learn from skeletons and preserved specimens. This print is a largely accurate representation of Leiden's theatre at this time, despite obvious use of artistic license. The advancement of anatomical knowledge is represented by the partially revealed dissected body. This contrasts with reminders of human mortality including the Latin mottos held up by skeletons and the Tree of Knowledge scene. (See also pages 6–7.)

© Royal College of Physicians [London]

Anatomy lessons

The Dutch Republic was a centre of trade and wealth in the seventeenth century which enabled the development of academic study, including medicine and anatomy, and also art. Painting became extraordinarily widespread, with large numbers of artists feeding an active market, including landscapes, portraiture and realistic genre scenes. One group of images depicts the anatomy lesson, including a lecturer, a body or skeleton, and a small audience. Nine such works, from 1603 onwards, portray the anatomy lecturers appointed by the Amsterdam Guild of Surgeons. While the Guild commissioned the work, all the men shown, including the audience, would have paid to be involved. In a culture where portraiture was as extensive as it was in the Dutch Golden Age, there was strong motivation to carry on in the same tradition.

The Anatomy Lesson of Dr Willem Röell, painted in 1728 by Cornelis Troost, does not depict an actual dissection or lesson. The anatomy teaching of the Amsterdam Guild of Surgeons took place in a dedicated theatre and was considered highly important for the education of apprentices. How-

The Anatomy Lesson of Dr Willem Röell, by Cornelis Troost, oil on canvas, 1728

This anatomy lesson was painted before Röell had given his first formal lecture as anatomy teacher to the Amsterdam Guild of Surgeons. The painting depicts the anatomist, and the surgeons who would have commissioned and paid the artist. As a result, the scene presents the participants as they would have wished to be seen, rather than showing the reality of anatomy teaching.

Archivart / Alamy Stock Photo

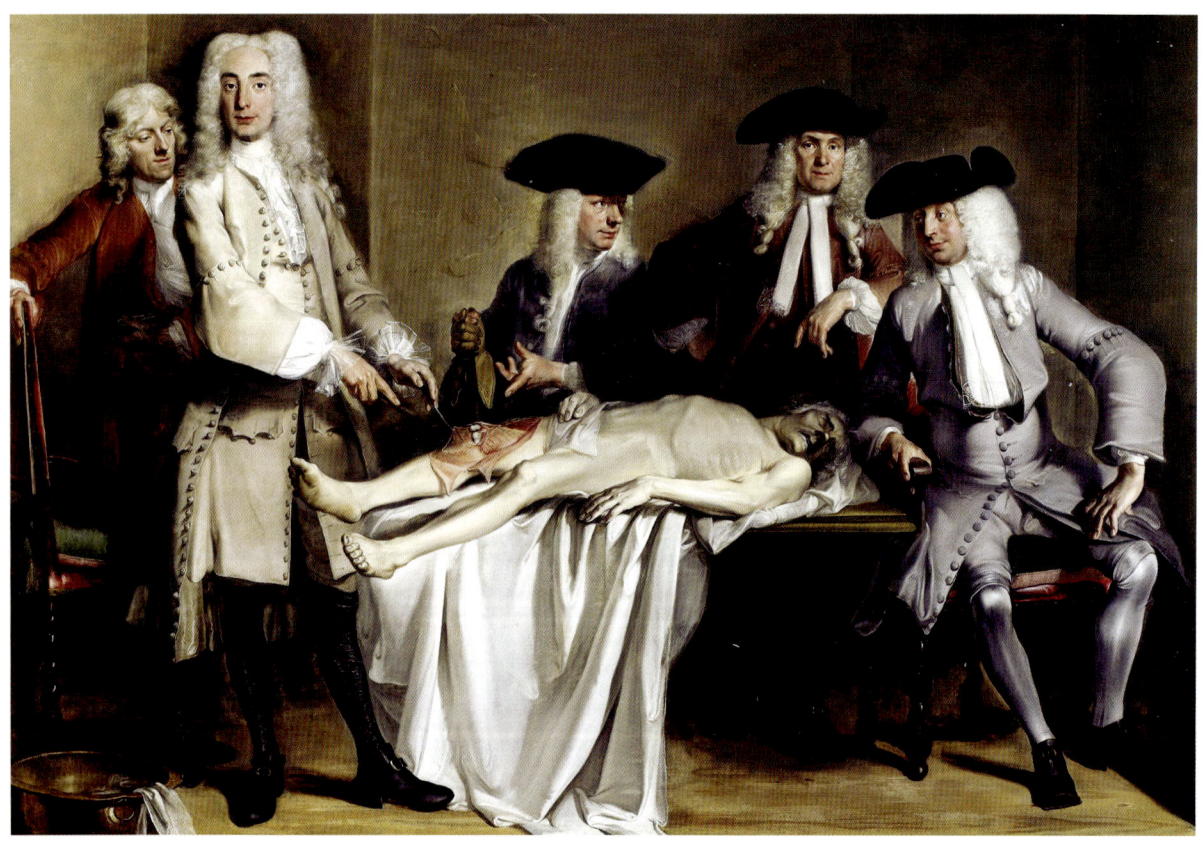

ever, these apprentices were normally consigned to the two back rows of the theatre, along with the paying public, while the front six rows thronged with members of the town council, physicians and surgeons, in that order. The anatomy lessons themselves were given by a university qualified physician.

At the time of this painting Röell, who qualified at Leiden in 1725, was not formally appointed *praelector anatomiae* (lecturer in anatomy) to the Guild. This took place in 1731 upon the death of his predecessor. In addition, he is not recorded as actually conducting a dissection for the Guild until January 1729, when he lectured for six days – with a two-day break for the weekend – on the body of an executed man identified by the initials 'NN' (execution records show that this was Timotheus Vernjou, a 22-year-old tailor).[6] The Guild itself noted down the date of execution wrong, implausibly stating a Sunday, perhaps the day the body arrived at the dissection room, and showing a lack of scrupulous detail about the identity of the body. Röell's first dissection and lesson therefore took place the year after Troost's painting which must therefore have been an early event of Röell's association with the Amsterdam Guild of Surgeons.

The painting does not tell us a great deal about how anatomy lessons were actually carried out. What it does reveal is how the sitters and the Amsterdam Guild of Surgeons wished to be perceived. In addition to Röell, the other men have been identified as three of the Governors of the Guild and the Guild Assistant, and the anatomy itself is relatively minimal, showing only a knee dissected. This is in keeping with other similar paintings, such as *The Anatomy Lesson of Dr Nicolaes Tulp*, painted almost a century earlier in 1632 by Rembrandt van Rijn, which shows the dissection of an arm. The arm in that case is portrayed with enough detail and accuracy to enable modern study of the anatomy revealed and analysis of the details. The minimalisation of the dissection could well be a feature of an artist's knowledge of anatomy. Painting a person would be a familiar and practised subject, and a live model could perhaps pose for it with the artist altering skin colour. However, rendering the technical details of a dissection correctly and at angles not shown in textbooks would have been more challenging, and with a greater need for a part-dissected body as a model which might not have been available at the right time. Painting a dissected body accurately would have taken more time and effort than a whole body – and the living sitters were paying for the painting and understandably prioritised themselves. These paintings were not intended as anatomical studies.

There may also have been concern to make the painting acceptable to a broader audience than those who would appreciate a detailed and realistic portrayal of dissection. The cloth guarding the body's modesty perhaps supports this concern. Of the anatomy lesson paintings that range in date from 1603 onwards, some show a skeleton, while others have a body with no visible dissection at all. The *Dissection of a Malefactor*, for example (see page 27), shows more anatomy than most, with a flayed '*écorché*' figure revealing the muscles under the skin. The skeleton and *écorché* figure would have been familiar anatomical representations to artists; however, the *écorché* in this particular painting lacks the detail and precision of an anatomical image and is perhaps more evocative than representational.

The *Dissection of a Malefactor* was painted by the Dutch artist Adriaen van der Groes and dated 1709 next to the instruments laid out on the body. It is thought that the central

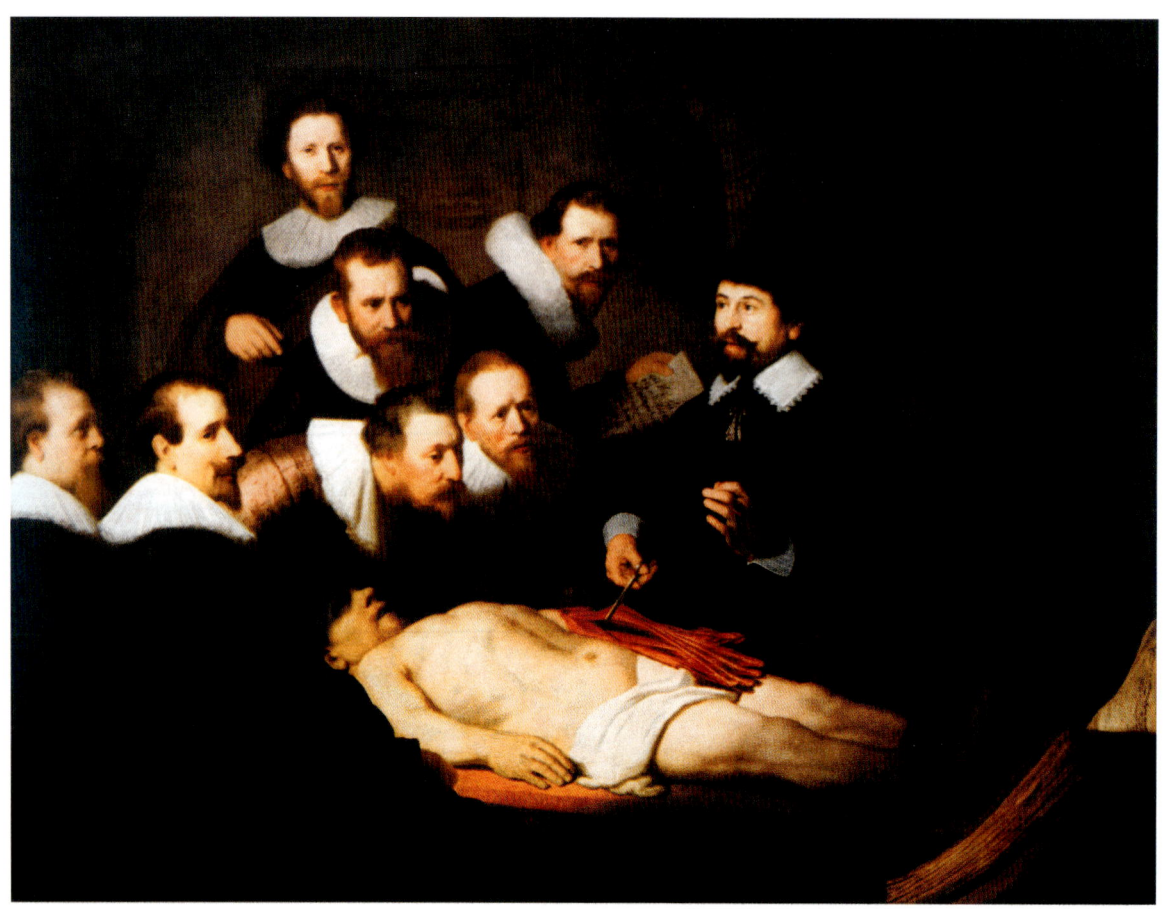

The Anatomy Lesson of Dr Nicolaes Tulp, by Rembrandt van Rijn, 1632

Adriaen Adriaensz was sentenced to death for stealing a winter coat. He was hanged and dissected in the only documented dissection of 1632 by the Amsterdam Guild of Surgeons. If this painting shows an actual body, it may be his.

Glasshouse Images / Alamy Stock Photo

figure is the Edinburgh surgeon and physician Archibald Pitcairne who had taught at the University of Leiden in the 1690s. This painting is interesting as it combines the genres of the anatomy lesson paintings of the Amsterdam Guild of Surgeons, which portray the dignitaries involved in relatively empty surroundings, with the allegorical imagery present in the prints of the Leiden anatomy theatre, with *memento mori* – reminders that you will die – and posed skeletons, *écorché* figure, banners, with a human skin draped over shelves at the top of the canvas.

The title, *Dissection of a Malefactor*, now given to this painting is unusual in emphasising the origin of the body as that of someone convicted of a crime, prioritising this above the name of the anatomist and the living people portrayed. Perhaps the latters' identities were so familiar and taken for granted that it was not deemed necessary to include them in the painting's title. The rope and the amputated right hand on the foreground floor of the painting invoke the punitive amputation of a thief's hand as well as hanging.

Dissection of a Malefactor, by Adriaen van der Groes, oil on canvas, 1709

This painting is thought to show the Scottish surgeon and physician Archibald Pitcairne, who taught in Leiden as well as Edinburgh. The symbolism of the rope around the dissected man's neck, and his amputated hand, indicated he had been executed, perhaps for stealing, which is reflected in the title given to the painting. A human skin hangs from the ceiling. The rope motif also appears around the necks of the skeleton and flayed figure at either side.

© Surgeons' Hall Museums, The Royal College of Surgeons of Edinburgh

Where did the bodies come from?

During this period, despite some variation between practitioners and places, there were patterns in the source of bodies dissected by anatomists. Those who were publicly dissected and acknowledged were executed people. They would usually be buried afterwards at the anatomists' expense, although some skeletons were kept. For research and private teaching, bodies were acquired in any way possible, as long as no one (of importance) would complain. People who died in hospital or hostels, and whose bodies were unclaimed by relatives or friends, might be quietly taken by doctors. Graveyards and battlefields were also places where the dead might be appropriated. The acquisition, and disposal, of bodies was a personal and private enterprise, with the student (or professor) of anatomy likely to venture out in person to source bodies for dissection.

Vesalius in Padua dissected bodies from local executions. His collaboration with a sympathetic judge (page 20) went so far as to include executions being scheduled to Vesalius' convenience, with the deaths of condemned people timed to his availability to dissect them. Also described above (page 18), Vesalius was requested by judicial authorities to confirm that a woman who was hanged had not been pregnant at the time, or at least was not far enough along in her pregnancy that the execution should be delayed. These were the bodies that were acknowledged in his public dissections.

For private anatomical study, however, Vesalius openly wrote in *De humani corporis fabrica* about acquiring bodies through other means, and encouraged his readers and students to secure them as and when they could. He himself appropriated a skeleton left hanging on the gallows, and found churchyards and vaults particularly easy sources of bones. The illuminated initials which decorate *De humani corporis fabrica* represent some of the processes of anatomical study, as carried out by naked *putti*. One of these – 'I' – shows a recently dead body being disinterred, and it has been suggested[7] that the Jewish cemetery in Padua may have been a source of fresh bodies due to the rapid burial after death and religious discrimination against Jews. That his book openly shows grave robbing was unlikely to cause difficulties, given the elite social classes who would have had access to his publication. The bodies taken for dissection were much more likely to come from the underprivileged and be considered a price worth paying for knowledge and medical advancement.

There is a possible exception, however, to the use of bodies of executed people for more open and publicised dissections. Execution records in Amsterdam noted a small number of bodies that were being sent to the anatomy rooms – usually one a year. This was typically in January as the cold would slow putrefaction and make a lengthy dissection less hideous. However, it is known that Willem Röell, as part of his remit as an anatomy teacher, also dissected the bodies of women for the education of midwives.

Boerhaave and Leiden

Leiden University was one of several universities in Europe where medicine and anatomy were studied. It was notable for its anatomy theatre, the second to be purpose-built after the one at Padua. The significance of Leiden as a medical school took a significant step forward from 1701 following the appointment of Dutch scholar Herman Boerhaave (1668–1738). In 1701, Boerhaave became Lector, or Reader, in the Institutes of Medicine (theory of medicine), but his duties rapidly increased. He offered private lessons in anatomy and chemistry, and in 1709 was appointed to the Professorship of Medicine and Botany, which included responsibility for the Botanic Gardens in Leiden. These roles go well together as the Botanic Gardens, among the oldest in the world, had been established in 1590 for the benefit of medical students. In addition, many prescriptions were based on plant or animal substances, either locally grown or sourced from further afield through colonial activities.

In 1714, Boerhaave was appointed as Professor of Clinical Medicine, responsible for teaching based on case studies around the dozen beds of the Leiden Infirmary. And in 1718, his great chemical expertise was recognised when he was appointed Professor of Chemistry. As the other components of prescriptions were mineral and chemical substances, study of chemistry at the university was also considered an adjunct of medicine. Boerhaave is further credited with being the first to use a thermometer to measure temperature in chemistry, and in medical diagnosis – although it did not become a common diagnostic tool until the late nineteenth century.

Despite holding three professorships simultaneously, Boerhaave did justice to his responsibilities and made Leiden into the foremost medical school of its day. As such, nearly 250 Scottish students studied under him, a disproportionate number given the population of Scotland. Among them were those who returned to Edinburgh and revolutionised the teaching of medicine there.

Bernhard Siegfried Albinus

One of Boerhaave's many influential former pupils was German-born Bernhard Siegfried Albinus (1697–1770), who taught anatomy and surgery at Leiden from 1719. Albinus continued the tradition of elaborately-illustrated anatomical works, clearly influenced by Vesalius. Working with the artist Jan Wandelaar (1690–1759), they produced the highly illustrated *Tabulae sceleti et musculorum corporis humani* (1749), incorporating plates of the skeleton and muscles of the human body – published mostly at Albinus' expense. The two men collaborated for decades, using technical drawing aids to ensure accuracy.

Dutch artists at the time are believed to have used the *camera obscura* – a drawing instrument that projected an image of the scene in front of it onto a flat surface – but artists appear to have maintained the mystery of their craft and individual skill by not advertising the fact. Albinus, however, describes the use of a screen or grid as a novel drawing technique to ensure correct proportions, despite such tools having been known about for centuries. It seems plausible that Wandelaar also used a *camera obscura,* or some other unrecorded drawing tools.

Notes

1. Jones-Lewis 2018.
2. Vasari 1568 (trans. 1907).
3. Hunter, W. in 1773, quoted in Clayton and Philo 2017, p. 27.
4. Leonardo da Vinci c.1507.
5. Blancken 1704.
6. Langenwagen 1766, p. 35. With thanks to Didi Trijp and B. T. Pierik for finding and translating the Dutch references.
7. Levine 2014.

Opposite: Line engraving by J. Wandelaar for B. S. Albinus' *Tabulae sceleti et musculorum corporis humani*, 1749

Albinus wrote of the decision to include the rhinoceros in the background of this anatomical image: 'On account of the rarity of the beast I thought that its form would be more pleasing than any other ornaments devised according to my own inclinations.'

Wellcome Collection. Attribution 4.0 International (CC-BY-4.0).

Background: Hydrostatic balance, probably made by Francis Hauksbee, England, c.1710–1733

This balance belonged to Andrew Plummer (1697–1756) of the University of Edinburgh. It was used in the research of medicinal spa waters.

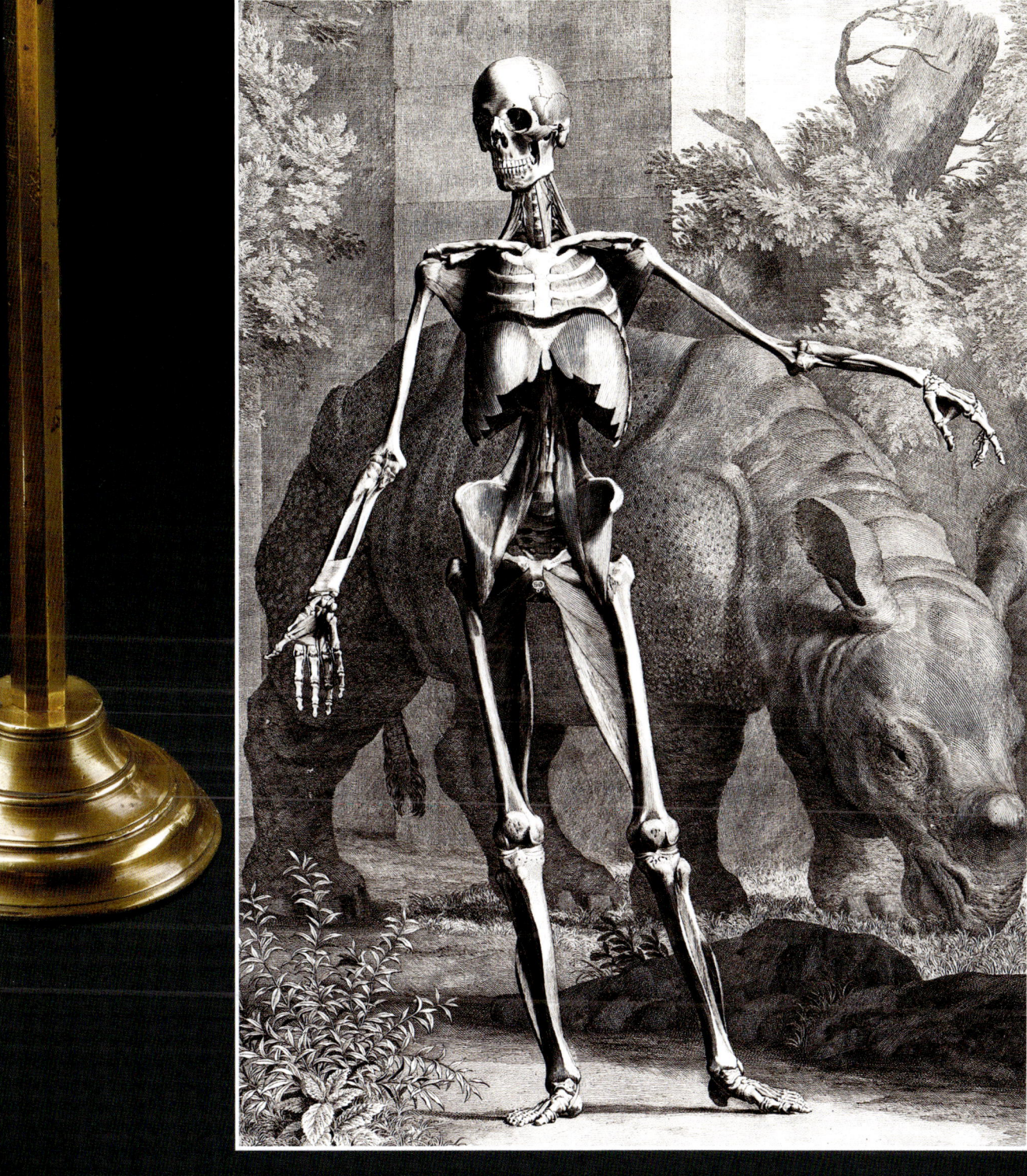

The Parliament Close and Public Characters Fifty Years Since, John Le Conte, Thomas Dobbie, after Sir David Wilkie
National Galleries of Scotland

Edinburgh and Anatomy

Scotland has a long reputation for valuing education for the wider benefits to society as well as increasing individual opportunities. In 1496 an act of the Scottish parliament was the first in the world to make schooling compulsory. Barons and landowners 'of substance' were required to send their eldest sons and heirs to grammar school from the age of eight or nine to learn Latin, and then to study arts and law for three years – 'through which justice may reign universally throughout the realm'. The demand was specifically laid on the social class likely to be responsible for justice and local government and able to afford the schooling, but did not infringe upon the liberty of the nobility to make their own decisions. Valuing education was not restricted to first sons and those for whom it was mandatory. While a child of school age would be past the risk of infant mortality, there was potential that a younger son would become heir following an elder brother's death. In 1505, when the Surgeons of Edinburgh incorporated into a craft guild, they began to establish academic standards for medical education in Edinburgh by requiring that all apprentices must be literate.

In 1583, Scotland's fourth university was opened in Edinburgh. Despite having around five times the population, England only had two (Oxford and Cambridge maintained their monopoly into the nineteenth century). The University of Edinburgh was established and run by the Edinburgh Town Council, which actively maintained control over the 'Town's College' and its staff appointments, curricula and examinations until an act of Parliament in 1858 removed this power. The Town Council's responsibility and interest in education was not confined to the University, but had broad local responsibility for the town, controlling crafts and professions through the guild and apprenticeship system. It made whatever appointments (not necessarily with remuneration) it saw fit, including teachers both at and outside the University, and the forerunners of public health officials. Personal contacts governed many of these appointments.

The Scottish Enlightenment, in which Edinburgh had a central place, was a time of intellectual exchange, advancement and openness to new ideas. Clubs and societies played a significant part of social and intellectual life in middle and upper class Edinburgh during the eighteenth century. They created a forum for the exchange of new ideas and allowed collaboration between men of different backgrounds and occupations. Membership was variously drawn from scholars, professionals and tradesmen, including the church and the military. The exchange between people in different fields and professions allowed men – and in a few cases women – to explore how ideas could translate from one topic to another. There were over two hundred clubs operating in Edinburgh throughout the century: formal and informal, literary and convivial. Medicine and anatomy flourished among the vibrant new ideas of Enlightenment Edinburgh.

Medical establishments

The development of medical education in eighteenth-century Edinburgh came from the alignment of several factors: the Town Council's interest in education; a practical desire to improve medical treatment and access; and the Scottish Enlightenment. The Town Council was keen to enable students to study medicine in Scotland rather than Europe and sought to improve the provision of medical education in Edinburgh from around 1700. It was therefore the Town Council that surgeon John Monro courted and influenced to establish medical teaching – and a key role for his son, Alexander, on his return from study at Leiden University. In 1720 the Town Council appointed the 22-year-old as Professor of Anatomy to the Town *and* University, whereas his predecessors had all been brought in specifically as professors to the Town rather than the University. It appears that John Monro was behind this detail of his son's appointment, and other future appointments, quietly working towards establishing a medical school at the University with the support of George Drummond, Lord Provost of the Town. Alexander Monro became the first of three generations of the same name to hold this post, monopolising the role for 126 years. Each professor in turn arranged through their connections for their son to succeed them, just as John Monro had done for Alexander. The three professors became known as Monro *primus*, *secundus* and *tertius* – first, second and third.

In 1726, the Town Council established the medical school at Edinburgh University, installing professors who had studied elsewhere, particularly at Leiden. Four doctors – John Rutherford, Andrew St Clair, John Innes and Andrew Plummer – were simultaneously appointed as Professors of Medicine and Chemistry, and the University was authorised to hold examinations in the subject without requiring external accreditation. All four of the new professors had studied at Leiden and worked closely together as a joint enterprise. Their appointments were also strongly supported by John Monro,

Medals and badges

Many clubs and societies met regularly at taverns and inns in the city's crowded Old Town, or in University buildings and libraries. Some awarded badges and medals to members upon admittance to the club or as prizes.

Left: Alexander Monro *primus* (1697–1767), by Allan Ramsay

Middle: Alexander Monro *secundus* (1733–1817), by William Scoular, 1820

Right: Alexander Monro *tertius* (1773–1864), by Sir John Watson Gordon

All images © Anatomical Museum, The University of Edinburgh

who may have written their speculative application to the Council asking that the roles be created for them. Rutherford taught the Practice of Medicine, later adding clinical lectures at the newly-established Royal Infirmary in Edinburgh, where those too poor to pay for care at home could get treatment and students could gain education and experience. From 1750 he was authorised by the Infirmary to hold autopsies on any of his patients who had died. St Clair led in teaching the Institutes of Medicine (theory and physiology), while Innes and Plummer concentrated on Chemistry and *Materia Medica* (drugs). In addition, they all maintained private medical practices and together operated a physic garden, growing medicinal plants, and an 'elaboratory' where pharmaceutical preparations were produced in large quantities in addition to teaching their preparation. Alexander Monro *primus* too was a practising surgeon.

These external activities, and the success of their lectures, were important because few professorships came with a salary. Rather, students paid individually for the classes they attended, and professors benefited directly from the popularity of their courses. While a few were compulsory for graduation, it was the topic that was mandatory, and attending a course at a different university, or held by another teacher accredited by one of the professional colleges, was accepted. Until about the mid-eighteenth century it was common for a medical student to study in several places and select a university to graduate from. Often the university that examined the student and gave the degree was not where much of the knowledge was acquired; and several universities were able to award medical degrees without having sufficient teaching available to qualify the student.

The new Edinburgh Medical School rapidly became a success, but was not the only avenue of medical instruction in Edinburgh. The longer established system of apprenticeship continued, especially amongst surgeons, although the Royal College of Physicians insisted on a university medical

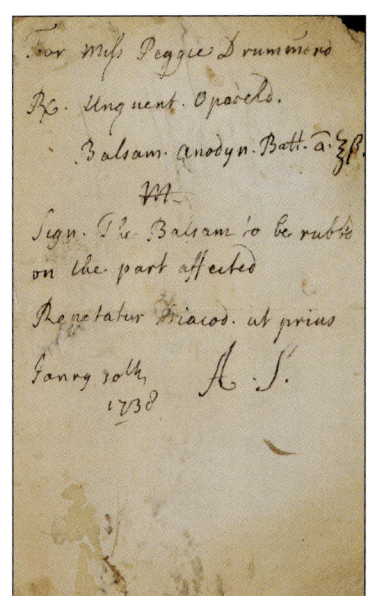

degree, from any university, as a condition for membership. But the various routes to qualifications were also not completely separate. Until 1725, Alexander Monro's classes actually took place at the Surgeons' Hall, as many of the attendees were apprenticed to surgeons and were not university students. The University's anatomy classes were in turn accredited by the Royal College of Surgeons as meeting their requirements for eligibility to stand examination. Classes were offered by University professors as well as teachers outside the university system. The city of Edinburgh and University together developed into a world-leading centre for medical education.

By the end of the eighteenth century, Edinburgh-taught doctors were finding careers all over the world, with far more Scottish doctors educated than decent employment available in Scotland. This further enhanced the fame and reputation of Edinburgh's medical teaching, attracting students from Scotland as well as British citizens from across the world.

Left: **Glass sample bottles for chemistry, belonging to Andrew Plummer**

Above: **Prescription written by Andrew St Clair for Miss Peggie Drummond, 30 January 1738**

Royal College of Physicians, Edinburgh. Image © National Museums Scotland.

Below: *Old Surgeons' Hall*, **built in 1697**, from a drawing by Paul Sandby, including figures from John Kay's *Portraits*

Midwifery

Also in 1726, the surgeon Joseph Gibson (*c.*1698–*c.*1739) was appointed by the Town Council as the world's first Professor of Midwifery. Unlike the other appointments, he was a Town rather than a University professor, because his expected students were female. Women were not admitted to the University and midwives were required to gain only a particular portion of mainly practical knowledge. His role also brought certification and control to a previously unregulated activity.

Midwifery would become far more medicalised over the eighteenth century with male physicians and anatomists developing practical knowledge to assist in difficult childbirths, asserting that knowledge over the female midwives. As midwifery qualifications were imposed, this was the one field in which women were admitted to formal medical teaching, but only to a certain level. A physician assisting in childbirth was not new, but a man midwife became an acknowledged medical specialism, claiming with a degree of justification that their education qualified them to provide better care.

Both infant and maternal morality were high, however, and caesarean operations had such a small chance of the mother's survival that the question of whether the mother's or the baby's life should be prioritised, when both were predicted to die without intervention, was a genuine debate.

Alexander Hamilton became Professor of Midwifery at the University of Edinburgh in 1783, resigning in 1800 in favour of his son, James. Both men made considerable advances to medical care during childbirth, sharing their improvements through the publication of textbooks. Alexander founded Edinburgh's Lying-In Hospital, where impoverished women could get maternity care, with the Hamiltons arranging the finance. The hospital admitted men and women students. In 1825 James went over the heads of the other medical professors to have the Town Council instruct that midwifery be taught as a compulsory rather than optional course for medical students at the University.

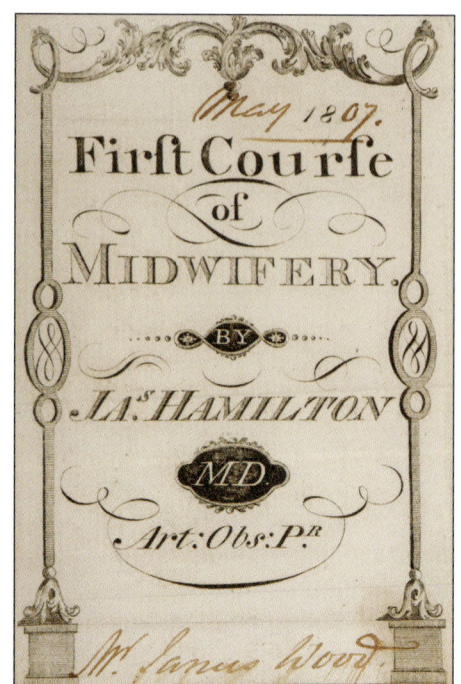

Above: James Hamilton midwifery class card for Mr James Wood, 1807

Below: An anatomical figure of a pregnant woman, ivory, Europe, *c.*1700–1750

The anatomy shown in this figure is not very accurate and would not have been useful for teaching medical students or midwives. Figures like this were perhaps used when explaining childbirth to newly-married couples or pregnant women. They could also have been elegant status symbols to decorate a doctor's room.

© Victoria and Albert Museum, London

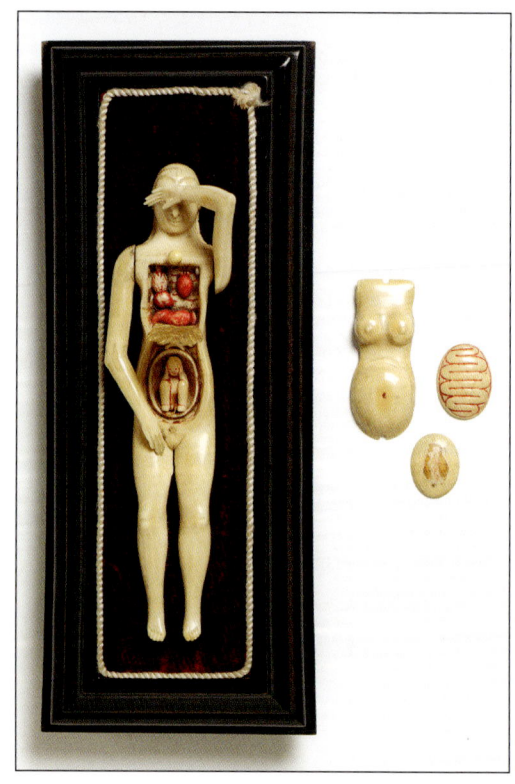

Medical control and standards

Many organisations in Edinburgh were invested in medical education and in controlling qualifications, and who could undertake which sorts of medical practice. There was a mix of competition and collaboration between the Town Council, the University professors and the professional colleges of the surgeons and physicians. Medical professionalism gradually developed along a number of different routes and coexisted with home remedies and traditional healers. This combined oversight and concentration of resources served to raise standards and status for medical education and practice in all related fields, including apothecaries and midwives.

The oldest medical establishment in Edinburgh was the Surgeons. Incorporated in 1505 as the Barbers and Surgeons, it restricted who could carry out surgery within Edinburgh and the education and standards that needed to be met. Outside the reach of a town's control, any individual would be able to practice surgery if they could find patients.

The combination of barbers and surgeons was a long-standing one, but not always a comfortable pairing as the professions developed. The surgeons treated the 'simple' barbers as second-class members, restricting their power within the guild and limiting where they could practice. This proved frustrating for the many who merely wanted to make a living cutting hair or making wigs, or customers who just wanted a hair cut without paying a busy and highly-trained surgeon to do it. In 1722 the Barbers and Surgeons officially parted company and separated into two organisations.

Seal of Cause of the Barber-Surgeons, 1580, copy of the 1505 Charter

Anatomical dissection was first approved in Edinburgh in 1505 in this charter which granted craft guild status to the Barber Surgeons of the city. In the charter, the Town Council approved the dissection of one condemned man a year, for the anatomical education of the apprentices. The Barber Surgeons would only accept literate boys as apprentices, as they began to establish educational standards for surgical practitioners. The charter was copied into this Town Council minute book in 1580 as it was relevant to the Council's discussions.

On loan from Edinburgh City Archives.
Image © National Museums Scotland.

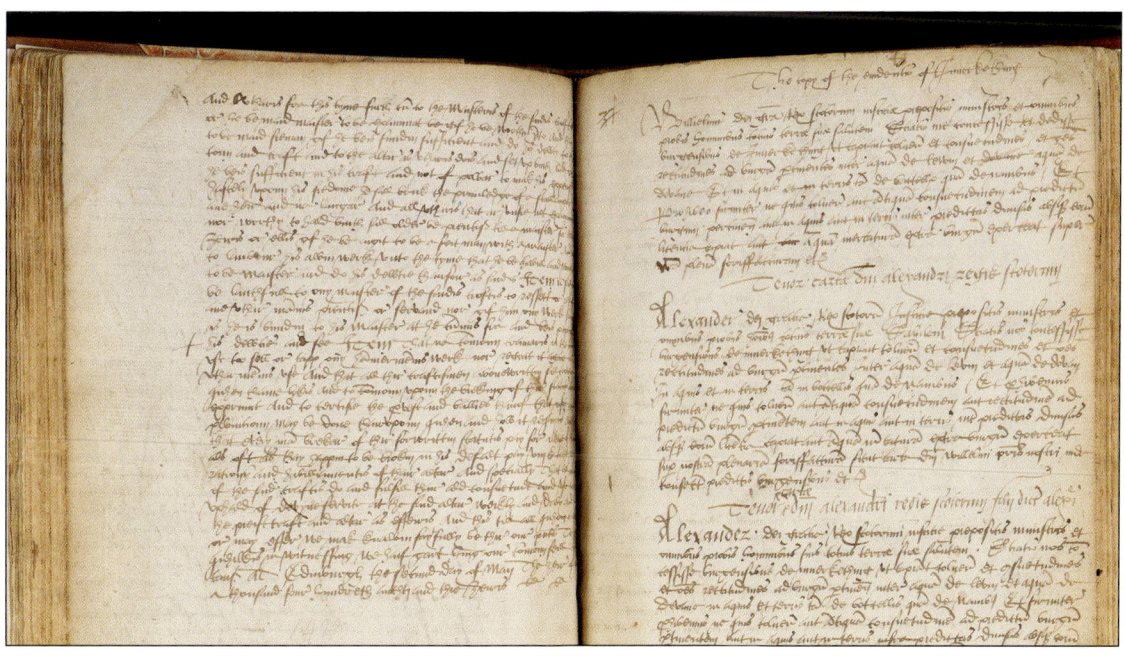

Medicine chest, 1746

This chest was supplied by Martin Eccles, a surgeon apothecary, who as an apprentice had attended Monro *primus*' first anatomy course in 1720. In 1742, Eccles and three of his apprentices were tried for 'violating the sepulchres of the dead, cutting and mangling their bodies', but they were not convicted (see page 45).

Another realm of medicine that was subject to disputes over control was the making up and selling of medicines by apothecaries. Many surgeons were surgeon apothecaries and the guild carefully watched the activities of apothecaries who were not members in case they trespassed into surgical activities. The work of apothecaries was also of interest to the Royal College of Physicians of Edinburgh, established in 1681. This College controlled the right to practice as a physician in Edinburgh, requiring applicants to already have a university medical degree. As the prescribers of 'physic', they had a strong interest in apothecaries and the College felt it should be in control of that profession in Edinburgh. In 1699, the Royal College of Physicians of Edinburgh published the *Pharmacopœia Collegii Regii Medicorum Edimburgensuim* – their pharmacy recipe book. This was an attempt to standardise medical treatment and to provide a reference work for physicians and apothecaries. It was also an unsuccessful bid to signal and establish the Physicians' position over apothecaries.

University-educated doctors, or physicians, were initially few in number and likely to be found only in Edinburgh and other larger cities. In most of Britain there were no restrictions on who could legally offer medical treatment. The care available was hugely dependent upon where the patient lived and their financial standing. A large proportion of the population did not have access to any individual with medical certification. Physicians could be, and often were, consulted by letter, but other treatments were required to be obtained locally.

Greater understanding of how the body worked did not rapidly translate into improved medical care for most people. Throughout the eighteenth and nineteenth centuries, home treatment remained an ordinary part of life, with medical recipes and nursing techniques collected alongside recipes for foods and cosmetics, often prepared by the women of the household. Charms, cheap and expensive, remained popular to treat and ward off sickness, with such misfortune generally assumed to have supernatural as well as natural causes. A charm at least had no harmful side-effects and modern medical research confirms how powerful the belief in a cure – the placebo effect – can be.

Consulting a qualified medical professional did not necessarily ensure better treatment, by modern standards, than home or folk medicine could provide. Many conditions simply had no effective treatment for physicians to prescribe. The Royal College of Physicians of Edinburgh's *Pharmacopœia* and surviving case notes and prescriptions, for example, contain many treatments that fell out of use for good reason as medical research revealed them to be ineffectual or toxic – these were eventually superseded. But as medical knowledge increased, improved treatments were gradually discovered, documented and made available to patients.

Charms and tokens

During the 17th and 18th centuries, medical professionals developed a greater understanding of how the human body worked, but this did not lead to better care for most people. Many people, rich and poor, used charms to ward off or to cure illness.

Some level of professional medical assistance was charitably provided for selected poor members of society, especially those expected to be cured and resume productive work. One of the acknowledged limitations to this treatment was the access of patients to decent food and rest. This was a motivating factor for the establishment of the Edinburgh Infirmary in 1729, following a fundraising appeal organised by the Royal College of Physicians, and by John Monro who lobbied hard for many medical causes. Its six beds immediately proved inadequate for the need, however, and the 228-bed Royal Infirmary was opened in 1741. It was both a charitable foundation for those who could not afford home treatment and a teaching hospital where students could learn directly from the cases of the patients.

Unqualified individuals did carry out surgical operations, sometimes for lack of a qualified medic or for reasons of cost. Indeed, there were some who developed considerable practical skill in a particular operation, such as pulling teeth, that might exceed the abilities of a broadly qualified practitioner with less experience in that specific area.

In 1741, the *Caledonian Mercury* newspaper reported that at the newly-established Royal Infirmary a man had been treated for an aneurism on his arm, caused by having his blood let by a gardener. Martin Eccles, the surgeon who treated him, thought it likely the arm would be saved, but the article was more an advertisement for seeking the services of a qualified surgeon and avoiding the dangers of those who were not – even if local and cheap.

> *Friday last John Low, who was taken into the Royal Infirmary, had the operation of the Aneurism performed upon him by Mr. MARTIN ECCLES Surgeon in Edinburgh. This Operation became necessary from his having the Artery of his Arm pricked by a Gardiner in the Country unskilled in Blood-letting, and to all Appearance is like to prove successful to the saving the poor Man's Arm. This and many other Instances of this Sort ought to render People very cautious, into whose Hands they commit themselves for the Performance of this nice, tho' seemingly slight, Operation of Blooding.*[1]

Anatomy and surgery

Bloodletting was a very common treatment – it certainly let the patient feel that something was being done. The surgeon, barber surgeon – or indeed gardener – carrying out the procedure needed to be able to find the vein or artery he required. Frequently, perhaps most conveniently, bloodletting is depicted on the arm, but other sites were used too, with various theories informing the practitioner which vein to recommend.

As well as locating a vein for bloodletting, other surgical operations required a knowledge of anatomy. Surgeons dealt with dislocations and broken bones which would have required an understanding of the skeleton and joints, as would occasional amputations. Before anaesthesia was invented, the speed of an operation was a key feature, and anatomical confidence enabled the surgeon to work rapidly. Surgeons were the medical practitioners who usually had responsibility for all external conditions such as wounds and bandaging, but also skin ulcers and boils. Many gained their experience working for the army, and the military's need for surgeons was behind some of the early national support for the profession, as well as a justification for supporting the practice of dissection and anatomy teaching.

Other surgeries were attempted, but they were dangerous and far from routine procedures. In 1742, for example, the aforementioned Martin Eccles found himself in the newspaper once again, this time for carrying out a successful hernia operation. This would have required anatomical knowledge of how body parts should be positioned in order to return them correctly to their place.

Left: Apothecary jars, 18th and early 19th century

Leeches were a standard part of medical treatment to remove blood from the body. Leech saliva contains protein which stops blood from clotting, allowing the leech to ingest blood up to ten times its weight.

Right (above and below): Pewter bleeding dishes, 18th century

Bloodletting was prescribed for any condition believed to be caused by a build-up of blood, ranging from indigestion to plague. A vein was pierced with a lancet and the blood was collected in a bowl.

Opposite: *A Perspective View of the Royal Infirmary, Edinburgh*, by Paul (Pierre) Fourdrinier, after Paul Sandby

The Royal Infirmary designed by William Adam opened in 1741 with 228 beds, providing care as well as teaching.

National Galleries of Scotland

Surgical instruments designed by John Weiss, razor maker to the King, 1823–1837

Specialised tools were used for different operations. This incomplete set contains Weiss' own design of amputation saw and a set of cranial instruments. Wooden handles and cloth-lined boxes like these fell out of use after sterilisation of surgical instruments became a concern.

We hear that about five or six Weeks ago, the Operation of the complete HERNIA was perform'd in this Place by Mr. MARTIN ECCLES Surgeon, upon Mr. Thompson Carpenter, who, notwithstanding the most dangerous Symptoms appear'd before the Operation, and that there was found prolapsed near a Yard of the Intestine, as also a large Portion of the Mesentery, is perfectly well recovered, goes abroad and carries on his Business.[2]

It is interesting that not only was the hernia operation considered newsworthy, but the paper seemed happy to publish the word 'mesentery' – the tissue that surrounds the intestines – without explanation. The report could have been copied verbatim from information provided by the surgeon, the equivalent of a press release being unaltered, or it may have indicated the level of anatomical knowledge expected of the newspaper-reading public by either the editor or the surgeon.

Hernia operations, however, were not always successful. One particular failure was recorded in the late eighteenth or early nineteenth century by the surgeon and anatomist Sir Charles Bell, who made a wax and plaster cast of his deceased patient to record the outcome –

> *... from an adult male who survived the operation of herniotomy during several days but without alleviation of symptoms. On the morning of his death repeated copious evacuation from the bowels occurred. On post mortem examination though the intestines showed some peritonitis there was no great intestinal distension. Though successfully reduced by operation the strangulated loop of intestine was black and gangrenous.*
>
> *In the right iliac fossa a coil of small intestine is much distended and its distal segment is black and gangrenous. The scrotum and right inguinal region show the incision of the recent herniotomy.*[3]

Other surgical operations that might be attempted included cutting for bladder stones, the removal of tumours, trepanning to treat pressure building up in the skull, caesarean delivery and tooth drawing. Anatomical knowledge was also expected of eye doctors. The diarist Samuel Pepys recorded on 3 July 1668 that he ...

> *met Mr. Pierce, the Surgeon, and Dr. Clerke, Waldron, Turberville, my physician for the eyes, and Lowre, to dissect several eyes of sheep and oxen, with great pleasure, and to my great information. But strange that this Turberville should be so great a man, and yet, to this day, had seen no eyes dissected, or but once, but desired this Dr. Lowre to give him the opportunity to see him dissect some.*[4]

The knowledge acquired from this dissection did not help at all in the treatment of Pepys' eyes. However, the surgical displacement of a cataract had been known for over two millennia in both Europe and India. This was a procedure that required anatomical knowledge of the eye and often resulted in blindness, although attempting it might be considered worthwhile when a severe cataract had already caused the sight to be lost.

Martin Eccles and Alexander Baxter

Martin Eccles, the surgeon apothecary, had attended the very first of Alexander Monro *primus'* anatomical classes in 1720. By the 1740s, however, his reputation was mixed and he had every good reason to desire the positive press coverage following the successful hernia operation. Only months before this, Eccles had been in the public eye for less favourable deeds. On 9 March 1742, the body of Alexander Baxter, buried a week before at the West Kirk of Edinburgh (St Cuthbert's), was found in a room beside Eccles' apothecary shop, triggering considerable rioting over several days. His shop was smashed, with Eccles and three of his apprentices arrested and tried for 'violating the sepulchres of the dead, cutting and mangling their bodies'. Although they were not convicted, two more of his apprentices who had absconded were assumed to have done so out of guilt.

This reflects the legal framework that surrounded human bodies at that time. A body could not be owned by any individual and, as such, having possession of one was not in itself a crime. What was illegal was the disruption of a grave – but with no evidence to link Eccles or his apprentices to the churchyard, the presence of the body was not proof enough to convict them. In addition, Alexander Baxter's body was probably stripped naked. While a body was not property and could not be stolen, the clothes in which it was dressed could be.

In the riots that followed the discovery of Alexander Baxter, the house of the West Kirk beadle, George Haldane, was demolished. Fragments of old coffins were found in the building, assumed to have come from graves that had been opened and the bodies sold for anatomical dissection. A sedan chair was burnt by the city executioner after a body was discovered inside it, and the two chairmen were banished from the city. A gardener, John Samuel, was widely believed to have been involved in digging up Baxter's body, but there was no proof. However, a few days later Samuel was discovered carrying the body of a child, named as Gaston Johnston, into Edinburgh at Potterrow Port. Again there was no proof to link him to the actual digging up of the boy, who had been buried nine miles away in Old Pentland. Despite this, Samuel was sentenced to be whipped and banished from the city. Perhaps, unlike the surgeon Martin Eccles and his apprentices, he could not plausibly claim to have merely bought the body. Thus he did not benefit from the support the authorities gave to medical professionals and their possession of bodies for dissection and study, if they could be acquired quietly and without scandal.

Edinburgh Town Guard halberd, 18th century

Demand and supply

In Edinburgh, as elsewhere, official support for medical education tacitly acknowledged that this required a supply of bodies and formal routes were unable to provide enough. Some bodies of executed people continued to be made available to anatomists; and in 1752 an act of parliament of the United Kingdom condemned murderers to have their bodies dissected after execution. This literally made dissection a punishment after death. But there were nowhere near enough murderers to meet the needs of anatomical education. Other bodies which might be dissected, with the agreement of the Town Council, included those of foundlings, abandoned babies, orphans who died in the care of the city, and prisoners who died in jail and were unclaimed by friends or relatives. In other words, those whom the authorities thought would provoke the least public outcry. Post-mortem investigation was also becoming more acceptable as it might provide information to the family about how or why their relative died, but anatomical dissection was not.

Even with the bodies of these mostly impoverished and poorly connected people, the official routes of supply were inadequate. Anatomists continued, as they had done for centuries, to take bodies where they could, without consent. They were dug up from graves, stolen before burial by false 'relatives', or simply appropriated when someone died alone in a hospital or other institution. Many were bought from grave robbers, body snatchers or 'resurrectionists'. Few people chose to sell the bodies of their relatives

Left: *The Old Town Guards*, by William Home Lizars, early 19th century, oil on panel

The Town Guard, and their successors, the police, did not patrol graveyards to watch for body snatchers. They simply tried to preserve the peace, stepping in when violence threatened between grave robbers or anatomists and the public. They would respond to specific information about a body that had been stolen and was now in an anatomist's room. Some anatomists hid bodies in concealed cupboards if they had advance warning of a police search.

The City of Edinburgh Council Museums & Galleries; City Art Centre

Below: Edinburgh Town Guard Lochaber axe, 18th century

directly – dissection was widely regarded with horror, as a violation of the body and upset of social customs. Some anatomists, more inured to dissection, believed buying of bodies could be increased. There are even a few reports of people approaching anatomists to try to sell the promise of their own body after death for money in life. However, this was seen as an insecure proposal by the anatomist, with no guarantee it would be followed through.

Official condemnation of these practices was weak. Grave robbing was known to supply the majority of the bodies that underpinned medical education. The authorities acknowledged anatomists would have bodies, and tolerated grave robbing so long as it did not lead to civic disturbance. Few laws were violated – and further laws were notably not established to criminalise further the activity, or the purchase and possession of bodies by anatomists. The disturbance of a churchyard, 'violation of the sepulchres', was a criminal act, but as a dead body was legally not property it could not be stolen. Therefore no action could be taken for merely having it, without a direct link to digging it up. The Town Guard and their successors, the police, would only take action if a specific individual's body was reported to be in the possession of an anatomist, to prevent vigilante action. However, many anatomists could hide a particular body if they had some prior warning.

After violent protests against body snatching in 1725, the Surgeons ordered that any apprentice found guilty of grave robbing should be dismissed immediately, and they had this decree published in a futile attempt to persuade the people of Edinburgh not to associate them with the practice. Of course, with the support of the authorities, being found guilty of grave

robbing was very rare, especially for a medical professional or student; and wealthier people, who had most access to the benefits of medical training, often considered grave robbing a price worth paying. After all, it would not be their bodies that would end up on the anatomist's table.

Inevitably the public took the protection of the dead into their own hands. As medical education in Edinburgh grew and brought an ever-greater demand for bodies for dissection, awareness of grave robbing and the need to protect graves increased too. Edinburgh had a comparatively large number of medical students for the size of the population, so demand for bodies was very high. Some graves were now protected physically, with the coffin locked inside a heavy iron mortsafe or watch house until it had decayed past the point of risk. As hiring a mortsafe was out of reach for many bereaved families, some arranged for watches to be kept on graveyards by those prepared to fight off the grave robbers. This led anatomists and students to purchase bodies from those prepared to dig them up, rather than venture out so often themselves.

The increased demand for bodies in Edinburgh led to them being shipped from places like Ireland, London and Liverpool. By the 1820s, a body fetched a very high price in Edinburgh, which made the shipping worthwhile. Likewise, it led to the bribery of those keeping watch over graves – to look the other way, or to offer information about burials with inadequate watch arranged. The money tempted people to risk discovery as grave robbers, falsely claim dead bodies 'of relatives' for burial from hospitals, lodging houses or poor houses, or to try to cheat the anatomists into buying a weighted box which did not contain a body.

Above: Mortsafe, cast iron, 1831

Relatives or friends could hire a heavy mortsafe to protect a coffin from grave robbers, but this was expensive. Unprotected new graves were not safe, even if they were far from cities and anatomy schools. Grave robbers were prepared to travel long distances to make money. This example is from Airth, between Edinburgh and Glasgow.

Below: Dark lantern, early 19th century

Dark lanterns, associated with grave robbing and burglary, shone light only when and where it was needed.

Left: *An Edinburgh Sedan Chair with Two Porters*, by David Allan, *c.*1785

National Galleries of Scotland

Below: Sedan chair, *c.*1780

Alexander Hamilton (page 38), Professor of Midwifery at the University of Edinburgh 1783–1800, and his son and successor James Hamilton, travelled around town in this sedan chair to attend women in labour. The chair was much more practical than a horse and carriage in the narrow dirty streets of the Old Town and reduced the risk of attack or robbery, particularly when visiting patients at night. This was one of the last private sedan chairs in Edinburgh.

John Dallas

In 1752, Helen Torrance and Jean Waldie, two women who lived just off Edinburgh's High Street, were hanged in the Grassmarket for stealing John Dallas, the eight- or nine-year-old son of a chairman (sedan chair carrier), and selling his dead body to a student. There was no direct evidence for the actual murder – presumed to have taken place between these provable events – and the defence argued unsuccessfully that selling a dead body was not illegal, and stealing a live child did not merit the death penalty.

This crime was thought to have grown out of a fraud gone wrong – Torrance and Waldie had intended to sell a weighted coffin to an unsuspecting anatomy student, but needed the false report of a death to provide the background for this enterprise. They could not get the co-operation they needed and the child's mother refused to pretend he had died. So, having promised a body they provided one, and were paid only two shillings and ten pence and another six pence for delivery.

William Hunter

Dr William Hunter (1718–1783) was an anatomist and man midwife from Lanarkshire, Scotland, who began his medical studies in Glasgow and Edinburgh and established a career in London. He rapidly developed his professional and social standing, becoming the foremost teacher of anatomy in London, a noted anatomical researcher and successful medical practitioner. Initially he intended to become a surgeon, but found the experience so unpleasant he abandoned that plan (according to his younger brother John, William had fainted while carrying out an operation). Instead he specialised in midwifery, becoming so successful that he was appointed physician to Queen Charlotte, consort of King George III, and delivered her children.

William's anatomical research extended to pregnant women and he used his knowledge to improve the care he offered in childbirth. This included the argument that many physicians were using forceps unnecessarily and how important it was to know when *not* to intervene.

As an anatomical teacher and researcher, William used a large number of bodies, enabling students to carry out their own dissections in what was then known as the 'Paris' manner of anatomy teaching. John Hunter joined him as an assistant, without any medical training. He soon proved adept at organising the supply of bodies for anatomical dissection, paying for those dug up out of graveyards or stolen before burial, even organising the targeting of specific people's bodies where they were thought to be anatomically interesting.

Left: Dr William Hunter (1718–1783) with *écorché* figure, attributed to George Michael Moser after Mason Chamberlin, enamelled copper miniature, 18th century

Right: Silver cup, attributed to Nicholas Crisp, London, 1760s

This was a gift from William Hunter's student class of 1761 and one of his most cherished possessions. He had been consulted on Queen Charlotte's first pregnancy and was hoping to be appointed as a Royal Physician. This new position would have allowed him to further his plan to open a National School of Anatomy and led him to consider closing his own school. But his pupils persuaded him against closure and presented the cup, depicting an anatomy class in progress, as a token of their appreciation.

Both images © The Hunterian, University of Glasgow

John went on to study anatomy and surgery, combining his knowledge of both, and became renowned for developing new surgical techniques based on his anatomical studies. He is said personally to have dissected two thousand human bodies during his career, as well as many animals. Both of the brothers collected anatomical specimens from their dissections, and from post-mortem investigations they were requested to carry out on patients who had died.

William requested that, after his own death, his body should be secured from grave robbers. John, on the other hand, specified that his diseased heart, and broken and healed Achilles tendon, should be added to his anatomical collection. This instruction was not carried out.

William Hunter's *écorchés*

William is recorded as having had at least three *écorché* casts made, which he used for his teaching, especially in lecturing anatomy to artists. What was and was not written down about the people whose bodies were cast is revealing of the attitudes that prevailed in elite circles at the time. The first cast was created around 1750, when William was lecturing at a painting academy in St Martin's Lane, instigated by the artist William Hogarth. In 1761, a small wax copy was made by Michael Sprang. Then in the 1760s, after Sprang's death, bronze casts were made of this wax. These small versions were convenient for study and acquired by both anatomists and artists. William Hunter was painted by Mason Chamberlin, lecturing with one of the small bronze casts in 1769.

A note, thought to be written (in the third person) by John Hunter, William's assistant at that time, reveals a little of the making process:

His brother who had the management of the dissections had eight men at once from Tyburn in the month of April. The Society was acquainted with it and they desired to come and chuse the best subject for such a purpose. When they had fix'd upon one, he was immediately sent to their apartments. As all this was done in a few hours after death, and as they had not become stiff, Dr Hunter conceived he might first be put into an attitude and allowed to

Écorché, by Edward Burch and Michael Henry Spang, 1760s

Bronze miniature copy of plaster cast *écorché*. This was the cast of the body of an unnamed man, hanged at Tyburn, London, in around 1750.

© The Hunterian, University of Glasgow

William Hunter lecturing at the Royal Academy of Arts, by Johann Zoffany, c.1772

William Hunter was the first Professor of Anatomy to the Royal Academy of Arts, founded in 1768. In this painting Hunter lectures to an audience of students, artists and members of the public, using as teaching aids a live model, skeleton and plaster figure made from a cast taken from the flayed corpse of a hanged man. Hunter bought hundreds of bodies from grave robbers for his research and medical teaching, but the bodies used in these more public lectures were those of executed people.

© Royal College of Physicians [London]

stiffen it, which was done, and when he became stif we all set to work by the next morning we had the external muscles all well exposed ready for making a mold from him, the cast of which is now in the Royal Academy.[5]

Unfortunately, the description of having eight bodies at once in April does not fit with the records of executions at Tyburn.[6] Only March 1750 or 1752, or May 1750, could have resulted in this number becoming available.

In 1771, for his teaching at the newly-established Royal Academy of Arts, William Hunter gave two lectures over the body of a thief, an unspecified one of four Dutch Jews hanged at Tyburn on 9 December, with their bodies delivered for dissection. Asher Weil, a surgeon, Levi Weil, Hyam Lazarus and Solomon Porter were all convicted following a burglary and the murder of a manservant, Joseph Slew. There were only two lectures given on the body so that it could be cast as an *écorché* while still fresh.

In 1776, William had another made, this time in the pose of a Roman statue possibly requested by the artist Augostino Carlini who did the casting. Although named *The Dying Gladiator* (now *The Dying Gaul*), it was also referred to as 'smugglerius' after the executed man from whom it was cast, said to have been a smuggler. From the 1776 May execution records, two smugglers were executed for the murder of a customs officer, Joseph Pierson, so this *écorché* could have been cast from either Thomas Henman or Benjamin Harley.

No one seems to have been sufficiently interested at the time to record the name of any of the three executed men whose muscular forms were duplicated in the *écorchés* mentioned here. However, they were all specified as executed criminals and in two cases their crimes were noted. This reinforces the notion that, despite many bodies being stolen for dissection and bought by the Hunters, their use was mostly kept behind closed medical doors. The more public presentation of bodies to artists used executed people, whose dissection could be openly discussed and their dissection noted as integrally linked to their crimes.

Both of the 1770s *écorchés* were cast from the bodies of men hanged for murder. Following the 1752 Murder Act, dissection was made part of the punishment for this crime and the bodies of convicted offenders were deemed the least controversial to make a display of. The emphasising of their sentences would forestall any protests or association with body snatching.

Collecting anatomy

Like the Hunters many anatomists collected preserved specimens for future teaching and study in three main categories: normal anatomy, morbid anatomy (or pathology), and comparative or animal anatomy.

Normal anatomy covered the healthy, except for being dead, human body with its usual variation. Skeletons used for student teaching were usually examples of this, and student and class dissections would mostly reveal normal anatomy. Diseased or injured organs from dissections or post-mortem investigations were particularly valued to show aspects that were less likely to be seen in any particular dissection, but important for medical education or study. They ranged from relatively common, such as arthritic joints or congested arteries, to rare tumours, congenital conditions and healed or fatal injuries. Sometimes casts were made where anatomical specimens could not be successfully preserved. Collections like these enabled students and qualified practitioners to observe for themselves conditions they might only occasionally come across and relate them to the case histories and symptoms experienced where these were known. Specimens taken from named, private patients ended up in collections, as well as parts of many anonymous bodies.

Animal anatomy was used as an easily available and unproblematic surrogate for human anatomy, but increasingly as a study of its own, as well as an opportunity to collect the exotic, unusual and different.

Walrus limb specimen, *Odobenus rosmarus*, from an articulated skeleton

Prepared by Robert and Frederick Knox.

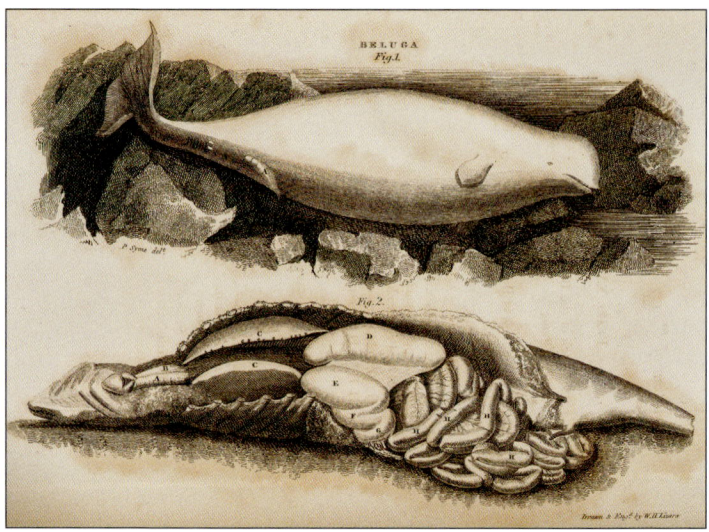

John Barclay

The Reverend John Barclay (1758–1826) was particularly noted for his comparative anatomy research and teaching in Edinburgh. He had been a church minister for about ten years before studying medicine and surgery. He then became a private anatomy teacher in Edinburgh, with his courses accredited by the Royal College of Surgeons of Edinburgh. His skills were so widely respected that there were discussions about creating the role of Professor of Comparative Anatomy at the University of Edinburgh specifically for him. This proposal divided the existing professors, many of whom, including Professor Monro *tertius,* feared losing income from student fees if the number of professors were to be increased.

In 1815, Barclay dissected a Beluga whale which had been killed in the Firth of Forth near Stirling. An account of this operation, described in an article published by the *Naturalist's Library*, gives a vivid impression of his lecturing style:

> *Never shall we forget the enthusiasm of the Doctor wading to his knees amongst the viscera of the great tenant of the deep, alternately cutting away, with his large and dexterous knife, and regaling his nostrils with copious infusions of snuff, while he pointed out, in his usual felicitous manner, the various contrasts or agreements of the forms of the viscera with those of other animals and of man.*[7]

John Barclay is reported to have said that he had no sense of smell, which may well have been a blessing as the whale was significantly decayed before it reached him.

Left: Rev. John Barclay (1758–1826), by John Syme, early 19th century

John Barclay was a leading anatomist and an anatomy teacher in Edinburgh. He particularly advanced comparative anatomy, preparing a vast collection of animal specimens. Among his students were William Dick, who founded Edinburgh's veterinary school with Barclay's encouragement, and Robert Knox.

National Galleries of Scotland

Right: 'Account of a Beluga or White whale, killed in the Frith of Forth', written by Dr Barclay and Mr Neill, *Memoirs of the Wernerian Natural History Society*, 1816

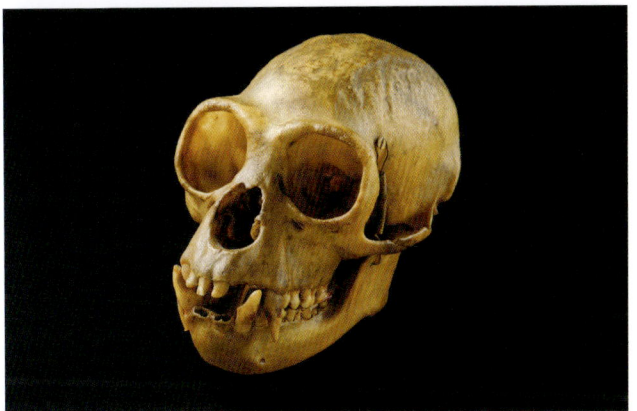

Robert Knox

Robert Knox (1791–1862) became the most successful anatomy teacher in Edinburgh – and the one who used the most bodies. He trained at the University of Edinburgh where, he later said, he failed his anatomy exams the first time, blaming this on Alexander Monro *tertius*' poor teaching. After attending classes by John Barclay, Knox passed his medical degree. No university record of this exists and it is probably one of many stories he told to enhance his standing with students and denigrate those he saw as rivals. After graduating, Knox spent eight years serving with the army, gaining experience as a surgeon and pursuing his own interest in the dissection of animals wherever he was sent.

On returning to Edinburgh in 1822, Knox persuaded the Royal College of Surgeons that they needed an organised museum – a collection of anatomical specimens – and that he should be appointed to care for it. This established him in Edinburgh with a position of anatomical and surgical standing from which to pursue his ambition to become Professor of Anatomy at the University of Edinburgh.

His reputation as a practical anatomist developed rapidly. In 1823 the second and third specimens of a duck-billed platypus arrived in Edinburgh, sent to the Professor of Natural History at the University. The platypus was

Left: Dr Robert Knox (1791–1862), by Augustin Edouart, cut paper, *c*.1830

French artist Augustin Edouart was the most prolific silhouette artist of his day, cutting the portraits of anyone with a claim to fame.

National Galleries of Scotland

Right (above and below): White-sided dolphin skull and gibbon skull

Comparative anatomy specimens prepared by Robert Knox and his brother Frederick Knox.

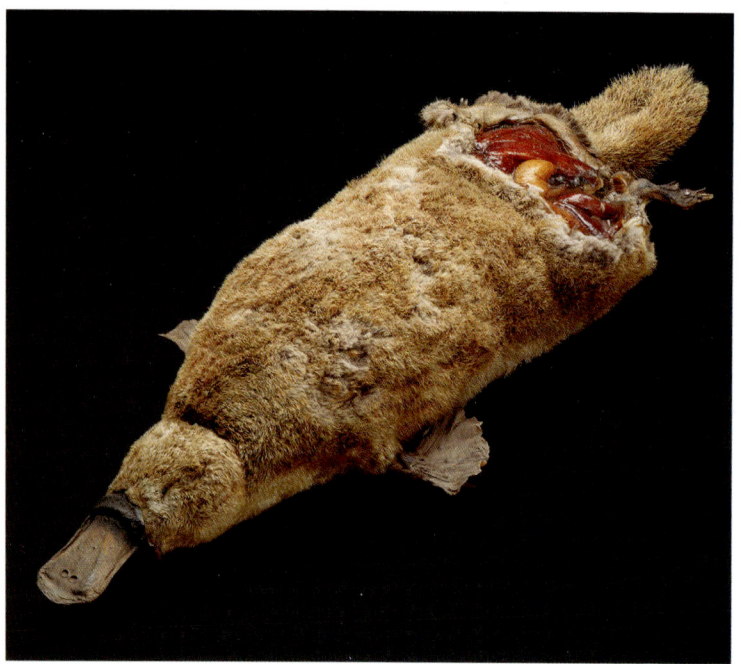
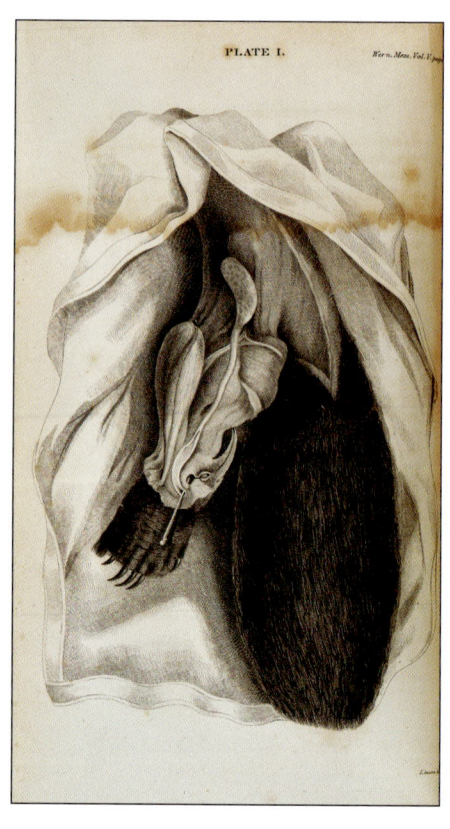

considered very strange and perplexing, although the animals were rather the worse for wear after a long sea journey preserved in alcohol, as was usual for foreign specimens. Robert Knox was asked to dissect one of the pair, recognising his skills as an anatomist.

In 1825, his former teacher John Barclay took Knox into partnership as teacher of anatomy, accredited by the Royal College of Surgeons. On the death of the former the following year, Knox took over the school at 10 Surgeons' Square.

Barclay had been a highly successful teacher and Knox built on this reputation. He succeeded in attracting even more students, cultivating his own image and popularity very carefully. He even maintained two houses: one where he lived and a separate residence for his wife who was not of the social standing to help advance his career. Knox was very popular with his students and employed a number of senior ones as assistants, as well as his brother Frederick to help with the Royal College of Surgeons' museum. His popularity, and that of his anatomy room, however, was mainly confined to his students. The American naturalist John Audubon visited Knox's anatomy school and noted that 'the sights were extremely disagreeable, many of them shocking beyond all I ever thought could be. I was glad to leave this charnel house and breathe again the salubrious atmosphere of the streets.'[8]

In 1831, a dead blue whale was discovered floating off the coast of Scotland and landed at North Berwick. Robert and Frederick purchased the whale, with Frederick taking three years to prepare the skeleton for display.

Left: Platypus, 1923

Such was Dr Knox's skill as an anatomist that he was asked by the Professor of Natural History at Edinburgh University to dissect and publish a description of one of the first three platypus specimens ever to arrive in the city. Two were brought there in 1823. Knox dissected one and this one was preserved mostly whole.

Right: Article on the dissection of a platypus by Robert Knox, *Memoirs of the Wernerian Natural History Society*, **Edinburgh, 1824**

For over a century, the same blue whale skeleton hung in what is now the National Museum of Scotland in Chambers Street, Edinburgh.

Dr Robert Knox was a prolific purchaser of bodies. In his lecture advertisements he blandly assured potential students that 'Arrangements have been made to secure as usual an ample supply of Anatomical Subjects'.[9]

Students

In the early 19th century, Edinburgh maintained its reputation as a centre of medical education. Alexander Monro *tertius* was Professor of Anatomy and Surgery at the University, and considered by many students to be old fashioned, and a less engaging lecturer than his father and grandfather before him. He was even known sometimes to read their lectures verbatim. Charles Darwin reported that he 'made his lectures in human anatomy as dull as he was himself'.[10] Monro *tertius* had the considerable prestige of the university professorship, but was in competition with the various private teachers for course fees. These were paid individually, with each student buying admission tickets to the classes they signed up for.

The case of Mathew Smith and Alexander Taylor

On 2 February 1807, 26-year-old Mathew Smith, a weaver and gardener, and Alexander Taylor, a 17-year-old apprentice to the surgeon James Kerr, were tried in Edinburgh's high court for a murder in Paisley, Renfrewshire. The victim was a two- or three-month-old baby girl, the daughter of Agnes Kelly. Mathew Smith was believed to be her father.

One evening Mathew Smith told Alexander Taylor he had a child whom he might have for dissection. He led him to a garden where the baby was lying on the ground. Smith was evidently surprised to find the infant was still alive and he immediately strangled and drowned her. He later claimed he had asked Taylor about the best way to commit the murder, but Taylor denied this. Taylor, however, did not deny that he witnessed the murder take place and then took the body back to his master James Kerr's house. Learning where the body had come from, the surgeon refused to let it remain in his house, but he did not report it to the authorities. The body was found two weeks later, in a cellar belonging to Smith's brother.

Mathew Smith was found guilty and executed on 11 March. The surgeon's apprentice, Alexander Taylor, was found not guilty of murder and no other charges were presented against him.

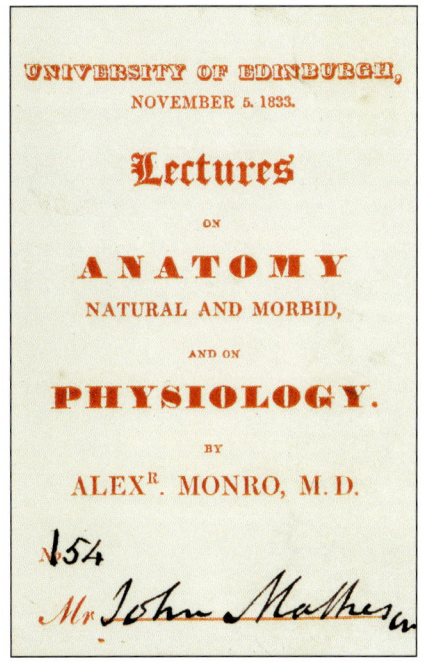

Below: Alexander Monro M.D., lecture card, 1833

A lecture card belonging to Mr John Mathewson, for the attendance of University of Edinburgh anatomy and physiology lectures.

Edinburgh city

Edinburgh in the 1820s was overcrowded, insanitary and unhealthy. Those who could afford to, lived outside the city centre or Old Town in the expanding suburbs. The anatomist Robert Knox, for example, resided at 4 Newington Place, just south of the Meadows, but still within walking distance of his anatomy rooms. Others had moved out of the crowded tenements into the New Town as it was gradually built from the 1760s. But this exodus into more salubrious dwellings was counteracted by the influx of economic migrants coming into the city in search of a living. Between 1791 and 1831 the population of Edinburgh and its neighbouring port, Leith, doubled from 80,000 to 160,000, and the building of homes did not keep pace. The creation of the New Town also had a detrimental effect on the hygiene of the Old Town as two of the three water pipes that supplied the city were diverted to the former, which even had the modern luxury of sewers. In the latter, sewage remained a nuisance and from 1822 cleaning was made the responsibility of the police. This was not the constabulary, but separately employed and very poorly-paid scavengers, many of whom were impoverished migrants from Ireland. Their job was to remove foul matter from the streets, where it was still thrown by householders each night. This mostly organic matter had value and much of it was either composted or carted outside the city for use as manure.

Over the eighteenth century, Edinburgh became a place of huge architectural contrast, with new spacious buildings being built for the wealthy. Organisations, including the Royal College of Physicians, Royal College of Surgeons of Edinburgh, and the University, found themselves in close proximity to the older buildings where a mix of social classes lived. Even into the mid-nineteenth century, many single room homes in tenements housed between five and fifteen people.

Notes

1. *Caledonian Mercury*, 4 May 1741.
2. *Caledonian Mercury*, 31 May 1742.
3. Charles Bell, Surgeons' Hall, The Royal College of Surgeons of Edinburgh.
4. Samuel Pepys 1668, diary entry from Friday 3 July.
5. Quoted in Postle (ed.) 2004, p. 57.
6. <http://www.capitalpunishmentuk.org/> [Mossop data]
7. Waterhouse 1841, p. 30.
8. Audubon, M. 1899, p. 174.
9. Archives and Manuscripts Collections, University of Edinburgh, EUA/IN1/ACU/A2/21/1.
10. Darwin 1892.

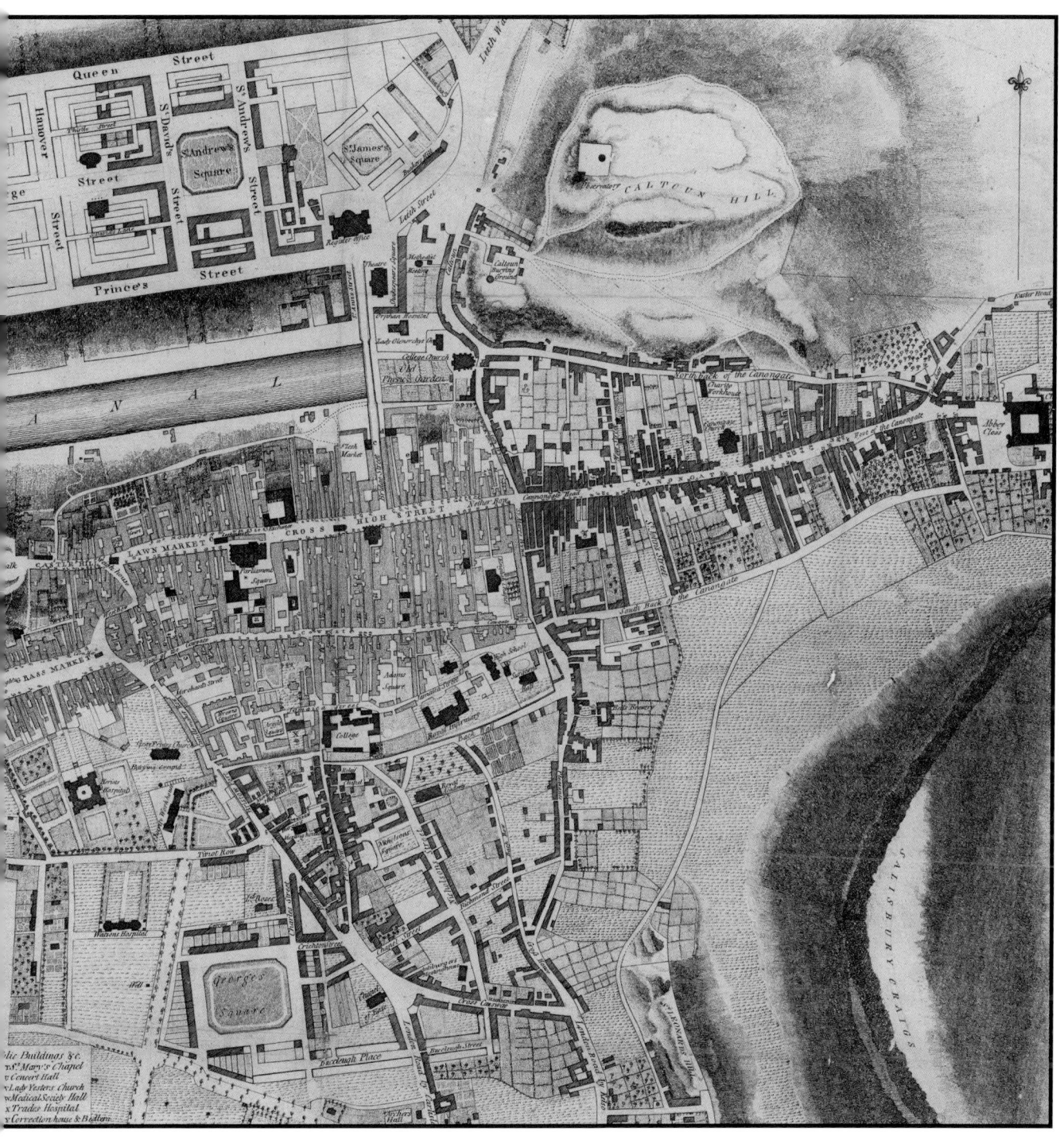

Map of Edinburgh city, John Ainslie (1745–1828), 1780: Reproduced with the permission of the National Library of Scotland (CC-BY) 4.0.
<https://maps.nls.uk/view/74400070>

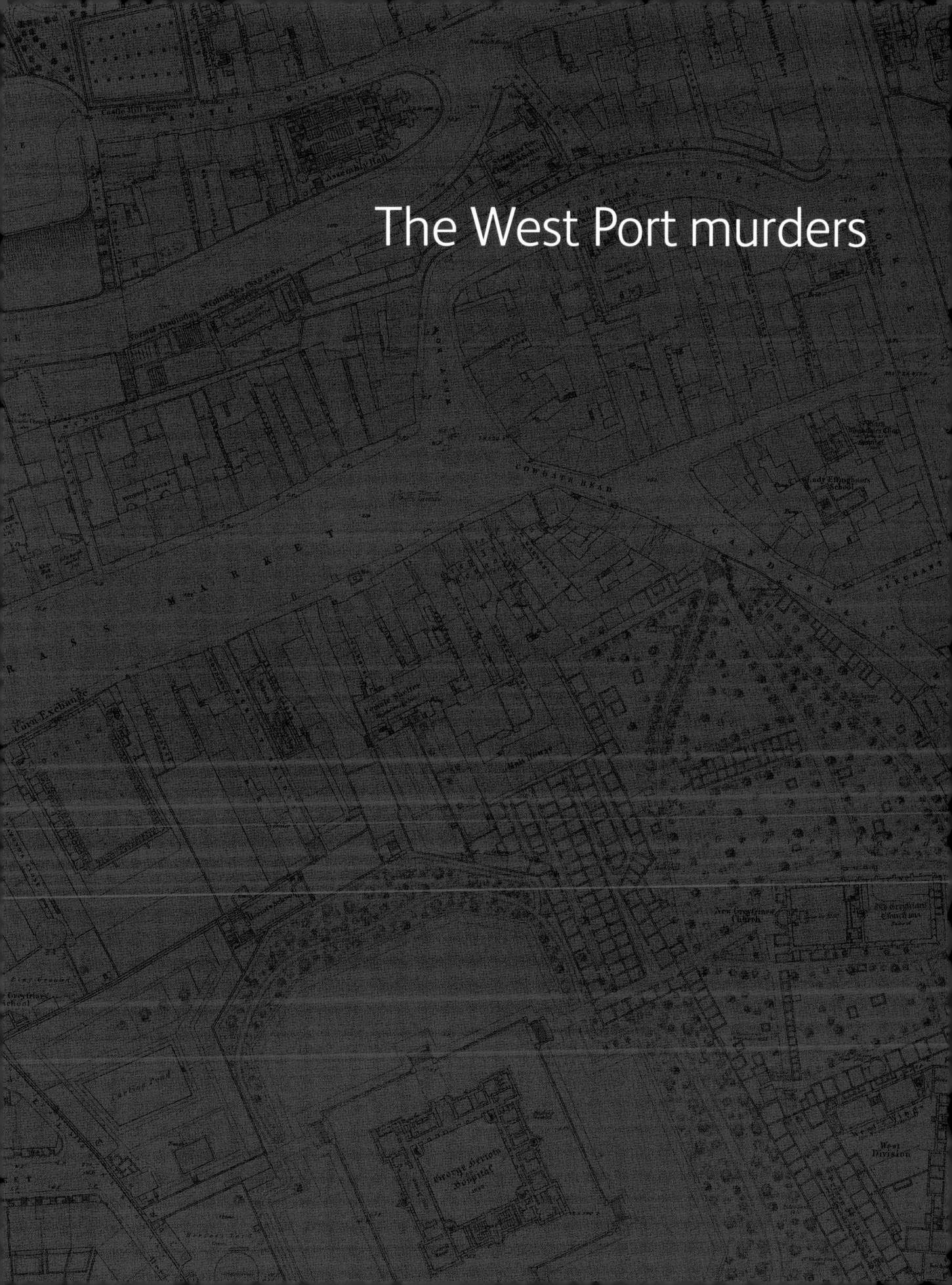

The West Port murders

On 28 January 1829, William Burke was hanged for murder in Edinburgh's Lawn Market. He had been found guilty of the murder of an Irish woman, Mary Docherty, whose body was discovered in his lodging in the West Port area of the city on 1 November 1828. Docherty's body was gone before the police were notified, but was recovered the next morning in the anatomy school of Dr Robert Knox.

Knox had bought the body for dissection from Burke and his companion, William Hare. When questioned, the doctor said that the freshness of the body had not raised any suspicions as he had bought a dozen or so in similar condition over the last year. This was initially seen as an implausible statement, as most bodies sold to anatomists were some days old. It was later realised that the statement was both true and chilling.

The murder of Mary Docherty was revealed to be one of a series of murders motivated by the price anatomists would pay for a human body. Yet after a difficult investigation and trial, Burke was the only person to be convicted.

At the time, these were referred to as the 'West Port murders', after the location where the perpetrators lived and all but one of the killings took place. They are now most often known and remembered as the 'Burke and Hare murders' after the two men involved – William Burke and William Hare. Burke's partner, Helen McDougal, and Hare's wife, Margaret Laird (or Hare), featured significantly in the story at the time, but less strongly in the memory and naming of the events.

As it unfolded, the story behind the West Port murders was very unclear. What was *not* known is just as important because it helps to explain how the investigation, and later the trial, developed. There was more suspicion and belief about events than actual proof – and suspicion and belief were not enough to secure a conviction and sentence of death in the court of law. Proof was absolutely vital.

With the benefit of hindsight, however, much of the story of the murders can be unravelled with a fair degree of confidence. This is based mostly on Burke's willingness to talk after he had been convicted of murder and sentenced to hang. At this point he is thought to have had little motivation to stray from being largely truthful. Thanks to the unfairness of Hare and his wife not being tried, Burke was likely to have shaded his story to portray his erstwhile friends in as poor a light as possible – and he was keen to protect McDougal.

Burke's confessions were compared at the time to Hare's now lost or destroyed confession and found to be in general agreement. Two separate transcripts of the trial of Burke and McDougal were published in the month after it took place and are another valuable resource, although the transcripts have a few issues of accuracy; additionally, the legal professionals, the counsel and judges, were invited to refine and doubtless to 'improve' their speeches for publication. There were also reprints of many newspaper articles, letters,

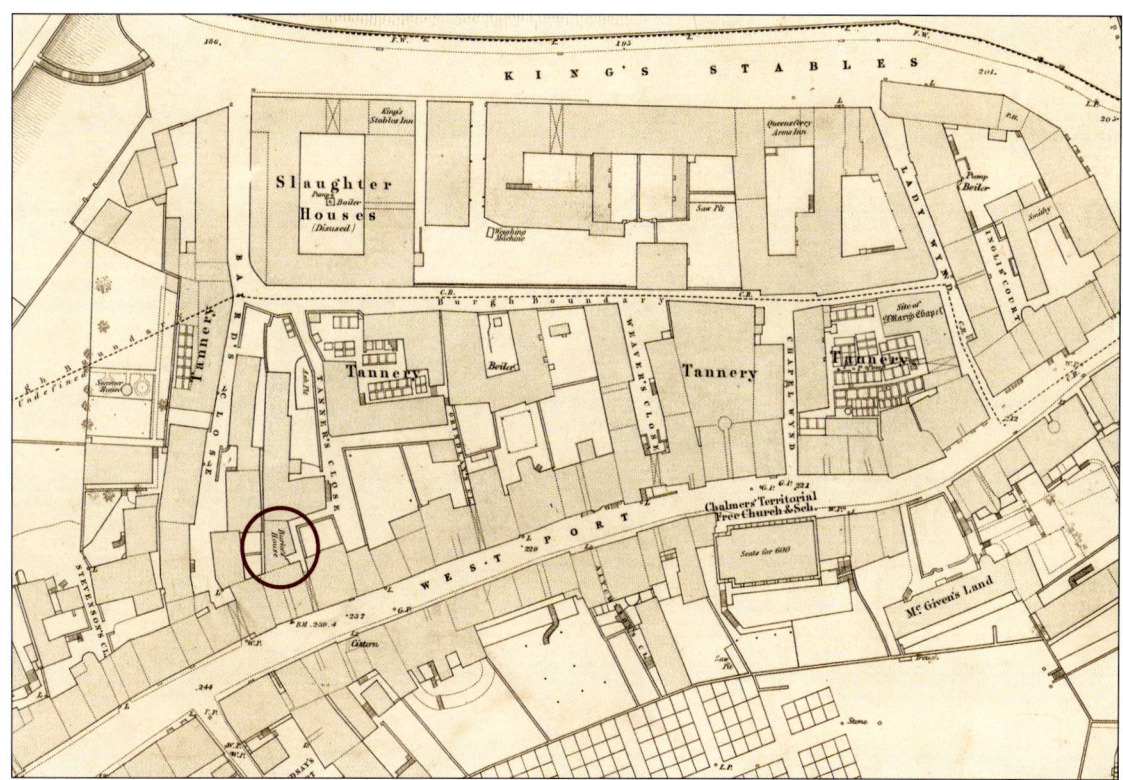

Edinburgh Ordnance Survey map, 1854

Detail showing the West Port area with the Hares' lodging house labelled as 'Burke's House' (ringed). The room Burke was renting at the time of his arrest is just above the 'P' of Port. (See also pages 60–61.)

Reproduced with the permission of the National Library of Scotland (CC-BY) 4.0. <https://maps.nls.uk/view/74415483>

Burke's confessions, and anything else a publisher might sell to take advantage of public interest. However, even with hindsight, there is much that remains unknown about the murders, particularly about some of those who were murdered. A number of these gaps have been filled in through stories and myths, and it can be hard to separate out truth repeated from personal accounts in the 1820s, from later invention or elaboration. Such stories and myths surrounding Burke and Hare continue to this day.

The Hares' lodging house

Among the incomers to the city of Edinburgh was a population of working-class Irish seeking employment. Margaret Laird was one such person, although it is not known when she had come over from Ireland. With her husband, an Irish labourer James Logue (or Log), she ran a slum lodging house in Tanner's Close off Wester Portsburgh, the road now called West Port. Despite being closely linked to Edinburgh and sharing resources, Portsburgh remained a separate burgh until 1856. In those days, Tanner's Close was true to its name, leading to one of several tanneries in the area conveniently located near the slaughterhouses and shambles at the west end of the Grassmarket, below Edinburgh Castle.

Margaret Laird was widowed in 1826 in her late twenties, and continued to run the lodging business. On 16 August that same year she married

one of her Irish lodgers, William Hare (or Hair), at the Catholic St Mary's Cathedral. Around January 1828 she gave birth to a baby.

There is surprisingly little known about William Hare, despite his infamy. We know he was born in Ireland, but not when. At his arrest he is reported to have said he was 21, but that he had come over from Ireland ten years previously in search of a living. He is variously noted to have been 25 years old or in his thirties. It is also not known for certain when, or where, he died.

Margaret Laird and William Hare were poor, but fairly secure financially in comparison to the itinerant workers or pensioners who found precarious shelter in the Tanner's Close lodging house. Such accommodation, according to a description of similar businesses, meant being 'boarded and bedded in an atmosphere of gin, brimstone, onions and disease, until their last penny be spent'.[1]

> *In Hare's house were eight beds for lodgers; they paid 3d. each and two, and sometimes three, slept in a bed; and during harvest they gave up their own bed when throng.*[2]

If the house was moderately full with two people in each bed, at threepence each, this would have brought in nearly a pound and ten shillings a week.

Hare and Laird's lodging house was in a small building with one other flat above, surrounded by taller tenements. It was almost certainly rented rather than owned. Contemporary accounts describe it as three rooms: two comparatively large with windows and fireplaces, and 'a small place eight feet or ten by four or five [which seemed] to have been formed by the staircase of another dwelling and the outer wall'.[3] The building was not demol-

Left and right: William Hare and his wife Margaret Laird

Drawn as they appeared in the witness box in court, December 1828. Margaret Laird's baby was ill with whooping cough.

William Hare (1792/1804–c.1858), artist unknown, after George Andrew Lutenor.

Margaret Laird (or Hare), by Thomas Clerk after George Andrew Lutenor.

National Galleries of Scotland

Left and right: William Burke and his partner Helen McDougal

Drawn during their trial, December 1828.

William Burke (1792–1829) and Helen McDougal, both by Thomas Clerk after George Andrew Lutenor.

National Galleries of Scotland

ished until 1902 and is shown in a few later images, both in photographs and drawings.

Hare continued to call himself a labourer and boatman, and maintained a cart and decrepit horse or donkey in a basement stable next door, used for hawking fish or crockery in exchange for old iron. The sloping land of the area gives some basements level access. He also had at least 21 pigs fed on scraps from wealthier households. These were possibly kept in the stable and roaming loose in the streets. In addition, the Hares employed a servant, Elizabeth Main – except during harvest when she could presumably earn more in labouring elsewhere (though this was when the house filled up with paying lodgers). After the Hares were arrested, Main was said to have stolen the pigs and appears to have escaped arrest for this.

Burke meets Hare

In early November 1827, another Irish labourer, William Burke, moved into the Hares' small room with his Scottish partner, Helen McDougal. Although sometimes called Mrs Burke by the neighbours, they were not married. Scots law recognised marriages by 'cohabitation with habit and repute', where couples could become legally married without a ceremony. Burke, however, had left a wife and two children behind in Ireland nearly a decade before, and so-called common law or irregular marriages could not be bigamous. Married women in Scotland often continued to be known by their own, maiden, name, but also accumulated those of their husbands as potential surnames.

Burke had been employed in the construction of the Edinburgh and Glasgow Union Canal, which runs from Edinburgh to join the Forth and Clyde Canal at Falkirk. The work was heavy and dangerous. After its completion in 1822, he made a living in a variety of occupations across Scotland and by 1827 was attempting to set up as a shoe-mender. Margaret Laird suggested that the small room in her lodging house would suit his needs for accommodation and workshop. This was probably sound business sense from Laird, seeking steady long-term lodgers. Burke later said that he already knew Laird: perhaps he and McDougal had lodged with Laird and her previous husband Logue at some point in the past, or Burke had met Logue during work on the Union Canal.

Burke's shoe-mending business was said to be reasonably successful and could earn up to a pound a week. Some shoes were brought to him to repair, but others he owned and McDougal hawked them on the streets.

The first body sold

Part of Burke's confession stated that 'an old pensioner, named Donald, lived in the house about Christmas 1827; he was in bad health, and died a short time before his quarter's pension was due – [...] he owed Hare £4; and a day or two after the pensioner's death, Hare proposed that his body should be sold to the doctors'.[4]

Around Christmas, less than two months after Burke and McDougal had taken up lodging with Hare and Laird, 'Donald', a long-term lodger in the house, passed away from dropsy. (Now called oedema, this is characterised by swelling from excessive fluid from heart failure or other cause.)

Burke's reference to 'Donald' in his confession is the only name recorded for him. He was described as an elderly man, in receipt of a pension. This indicated that he was most likely an old soldier as army pensions were the ones most commonly established at the time. However, military records do not survive in sufficient detail to identify if there was an old soldier by the name of Donald who stopped drawing his pension at this time.

Robert Knox's former student and rather partisan biographer, Henry Lonsdale, gives a earlier date of 29 November for Donald's death. However, during the murder investigation, Knox was so unforthcoming with records that it seems unlikely he would have revealed any precise information at a later date. 'About Christmas' also fits more closely the description of being shortly before a quarterly pension would be paid.

Hare complained to Burke that Donald owed him £4, which would not now be recovered. The debt seems unlikely to have been accumulated merely through a place to sleep at 3d a night, as this would imply a debt stretching back over ten months. Nor did Hare appear to have savings enough to lend a sizeable sum. Perhaps Donald had agreed to pay his entire pension to the Hares in return for food and lodging.

It was generally known at that time that a grave needed to be watched to prevent body snatchers making off with the body. Hare, Burke, and everyone in Edinburgh, were well aware there was money to be made from the sale of bodies. So the two men replaced the deceased man in the coffin with a weight of bark from the nearby tannery and set out to find a doctor to buy the body, with little attempt to disguise what they were up to. The protection of the bodies of the dead was undertaken specifically by family and friends of the deceased, but there was less care about the bodies of strangers.

It is perhaps unsurprising that Burke and Hare were directed to the anatomist with the most students and known for the greatest use of bodies – Dr Robert Knox at 10 Surgeons' Square. There they met with some of the doctor's senior students employed as his assistants, who were clearly more experienced at buying bodies than the two men were at selling them. Later than evening, Burke and Hare carried the body to the anatomy rooms in a sack, only to discover that it was expected bodies would be delivered unclothed. This probably related to a point of law where a body has no owner so it cannot be stolen, but clothing could be.

Knox looked at the body and authorised the payment of £7.10/- without question. Would any other anatomist have enquired more deeply? Probably not. The swelling of dropsy would have been immediately obvious, indicating the illness and likely cause of death, and almost the entire supply of bodies for anatomy involved purchase from financially-motivated individuals who had some secret or underhand aspect around the acquisition of the body being sold. This transaction was in fact exactly what it appeared to be – a man who had died from dropsy and who had no family and friends to ensure that his body was buried safely and did not fall into hands of body snatchers. However, the price Knox paid was a bit low when compared to later transactions, perhaps due to the added risk of buying from unknown suppliers with whom he had not previously dealt. The payment, however, was extremely significant to the sellers. As Burke said in his confession after conviction, 'getting that high price made them try murdering for subjects'.[5]

Knox said he would be pleased to buy more bodies from Burke and Hare in the future. He even had them informed when medical colleagues told him about a poor person who had died without family around them, as Donald had done. Knox thought people might be persuaded to sell such bodies to Burke and Hare, who would then deliver them to him. But Knox never received any such bodies from Burke and Hare – instead, they brought others.

And this is how the matter stood – with Knox purchasing bodies for his anatomy school from these particular suppliers, with no more secrecy than any other source, until 1 November 1828 when the affair came to the attention of the public and the authorities.

Murder is suspected

In the middle of 1828, Burke and McDougal moved from the Hares' lodging house to stay with a relative and her husband, the Brogans, in their single room dwelling, only 7 feet 5 inches by 16 feet 2 inches (2.26 m x 4.93 m) with a window at one end and an open fireplace for cooking in one wall. It was situated on the lower ground floor, level with the cellars in the front of the building, one of the less desirable parts of a tenement, close to the stink of the streets. But the window looked out onto the back courtyard, rather than the busy West Port.

The Brogans, Burke and McDougal had two neighbours on their floor: Ann Connoway and her husband John, and Janet Law. Mrs Law appears to have been a mangle keeper, a woman who made an income through ownership of a mangle which washerwomen and

 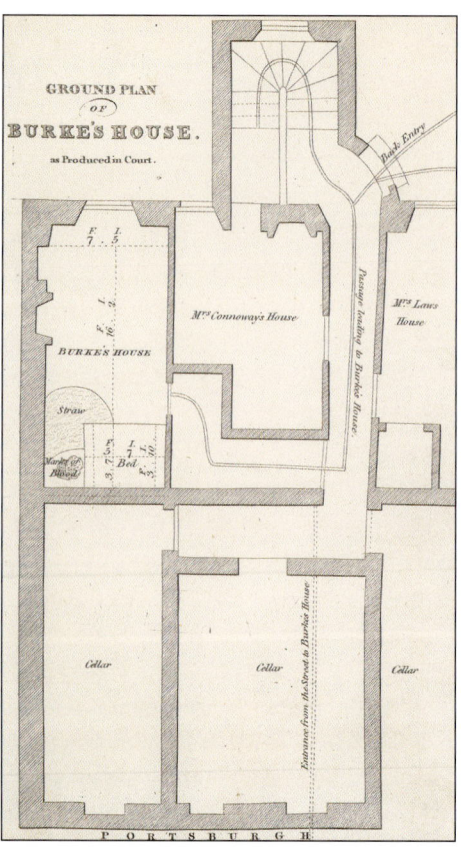

Left and right: William Burke's lodging

Engraving of the window (marked A, ringed) of his lodging and its back entrance (marked B), and a floor plan as presented in court of his single room in which four people were murdered.

William Burke's House, by Walter Geikie.

'Ground Plan of William Burke's House, by Thomas Clerk.

National Galleries of Scotland

other women would pay to use. This would have been what is now called a box mangle, a large and cumbersome device for pressing and flattening cloth as an alternative to ironing, which must have taken up a considerable part of her room.

As well as a lack of space and privacy, the people in the area had few possessions of their own. Mrs Law commented that Helen McDougal was in the habit of borrowing things from neighbours, noting, 'Mrs Burke came into my house to borrow a pair of bellows […] that was about eight. She afterwards returned about nine to borrow a dram glass.'[6] McDougal herself said that the coarse linen sheet on the bed she shared with Burke belonged 'to a William McKinn, from whom [she] got a loan of it'.[7] In return, both Mrs Law and Mrs Connoway found cause to enter the McDougal and Burke household several times a day. The door had no lock, just a bolt on the inside – but one advantage of so many people living close together is that someone was likely to be around to watch the door.

It was in this crowded and communally-used room that a man called James Gray reported to the police that, at about six o'clock in the early evening of 1 November 1828, he had witnessed his wife Ann discover a dead body hidden under the bed and a heap of straw. But a short time later, there was no body to be found anywhere in the room or adjacent cellars. With so many people around, surely someone had noticed something.

Towards the end of October 1828, the Brogans had gone away, leaving

Burke and McDougal in the home, where they were joined within a couple of days by James and Ann Gray as lodgers, with their young child. Ann was the daughter of Helen McDougal's former married partner, and the Grays found Burke and McDougal kind and welcoming. On Hallowe'en morning, 31 October, Burke brought another woman to the house. He had met her at one of the many spirit dealers in the area well patronised by himself. Mary Docherty, also known as Mrs Campbell, was Irish and appeared to be in her forties. She had come over from Glasgow to see her son in Edinburgh, only to be rejected by him.

Burke had struck up a conversation with Docherty, asking whether they might be related. This led to an invite to join his household to celebrate Hallowe'en. Burke told the Grays they would need to leave that night and arranged accommodation at the Hares' lodging house. When they returned for breakfast the next morning, Burke was behaving in an irritable and unusual manner, particularly around the bed and straw in part of the room. McDougal, meanwhile, was 'in bed unwell, in consequence of drink which she had had'.[8] John Brogan, the son of the former tenants, was present, as well as Mrs Law and Mrs Connoway who were visiting. Mary Docherty, however, had apparently left.

Burke went out, asking Brogan to stay in the house. But by evening, when the Grays were finally alone in the room, Ann took the opportunity to investigate the bed and straw around which Burke had been behaving so oddly earlier. He had seemed to be keeping her away from the area, despite her child's clothing being there, and the potatoes for dinner. Underneath the straw, Ann discovered Mary Docherty's naked body.

James Gray immediately packed their belongings, while Ann replaced the straw. On leaving the house they met McDougal on the stairs. When challenged, McDougal said that Mary Docherty had died in an accident and she tried to bribe the couple to keep quiet about the body. Margaret Laird, meeting with the Grays on the street outside, coaxed them into a public house rather than discuss the matter noisily in the street. She left them there. But James Gray remained unmoved by either the explanation or the bribe and went to the police later that evening – after prudently collecting a few last bits of clothing from Hare's lodging house where they had spent the previous night.

When a police officer arrived at Burke's room between about seven and eight on the evening of Saturday 1 November, there was no body to be found. This was hardly surprising. In trying to persuade the Grays not to report the body, McDougal and Laird said it would be gone and sold before the authorities could see it. There were, however, some bloodstains that required explanation. McDougal said that 'a woman had lain there' about a fortnight ago and the bedding had not been washed since.[9] Unwashed bedding was not unusual, even with period blood, but these stains looked less than two weeks old. Burke and McDougal were also inconsistent in their answers as to where Mary Docherty had gone. Despite having had time to prepare, Burke said she had left at seven in the morning, while McDougal stated seven the night before. This was suspicious.

The officer made an excuse to get Burke and McDougal to the police office and returned to the room with the police surgeon and superintendent to search for more clues. They found additional relatively fresh bloodstains, some mixed with saliva, and some of Mary Docherty's clothes.

At seven the next morning, Sunday 2 November, the police turned up at a David Paterson's doorstep, at the place where he lived with his mother and sister, also in the West

Port. Paterson was employed by the anatomist Dr Knox and termed himself the 'Keeper of the museum of Robert Knox', although in a janitorial rather than intellectual capacity: 'nothing more than a menial servant, hired by the week at 7s[hillings], and dismissible at pleasure',[10] according to Knox's anatomical assistants. That the police went straight to Paterson and not to the anatomy rooms in Surgeons' Square seems likely to have been the result of their enquiries rather than speculation about where a body might have vanished to. Both Paterson and his 15-year-old sister had called at Burke's house during the 24 hours before Gray reported the body, and McDougal in her attempt to bribe Gray to silence had made it clear the body was to be sold. Paterson took the police to the cellar at Surgeons' Square where the body of a short woman, just as described, was discovered crammed into the same tea chest in which she had been carried there.

The investigation

Now began the challenge to unravel what had happened to this woman. Mary Docherty had been healthy enough to make her own way from Glasgow to Edinburgh to visit her son on 30 October, and then to talk of leaving town on 31 October and also mention trying to find her son again. Yet on the morning of 2 November, her dead body was identified at the police station, brought there from Robert Knox's anatomy rooms in Surgeons' Square. Had she been murdered? Or had she died through misadventure and her body opportunistically sold? Forensic science of that time was of little help. Alexander Black, the experienced police surgeon, noted that she appeared to have been smothered or suffocated. Unfortunately for the prosecution's case, Black was prepared to swear there was no evidence to suggest this was *not* accidental. He had, he said, on one night dealt with no fewer than six people who had been brought in from the streets having suffocated from drinking to excess and passing out face down. He would scrupulously report that while he personally believed she had died by violence, he could neither medically nor scientifically prove it beyond doubt.

To modern readers it may seem improbable that six people should suffocate from drink during a single night in Edinburgh. Alcohol may not have been the sole cause, however; it could have been taken in combination with opium easily acquired from any apothecary. Whatever the medical cause of death, it emphasises the mortality rates of the time.

The body was additionally examined by the eminent surgeon William Newbigging, and by Dr Robert Christison, Professor of Medical Jurisprudence (medicine as it relates to law, including the forerunners of forensics) at the University of Edinburgh. In 1828, it was still the only professorship on this topic. Both men concurred that death appeared to have been caused by suffocation, but could not prove that it was murder and not an accident.

Burke, McDougal, Hare and Laird were repeatedly interviewed, all telling contradictory stories about the night in question. Nonetheless, a horrifying suspicion was beginning to grow that this was not just a single murder investigation but much more than that. Mary Docherty was not the first fresh, unburied body that Burke and Hare had sold to Knox as a subject for dissection. As the story spread and rumours circulated, huge public pressure began to build. Had this been the fate of other individuals who had disappeared in Edinburgh during recent times?

Most talked about among the missing folk were James Wilson and Mary Paterson. Wilson was a familiar street figure around Edinburgh, known as Daft Jamie. Had anyone seen him in the last couple of weeks? And were those definitely his clothes being worn by Burke's nephews? Janet Brown had been looking for her friend Mary Paterson since she vanished in April from the house of Constantine Burke (William's brother), where they had been drinking with Burke and Hare. Had the bodies of Jamie Wilson and Mary Paterson also ended up on the anatomy tables? Yet still there was very little concrete evidence.

Dr Knox said he believed, as far as Burke and Hare were concerned, that he had been dealing with body snatchers or buyers – people who acquired deceased bodies before burial, spirited away if they died on the street, or purchased from landlords if they died in their houses (just as army pensioner Donald had done). It was an unavoidable fact that death occurred – and every subject in the anatomy rooms had been alive not long before, whether they had been dug up by grave robbers or had never made it to the cemetery.

Burke, Hare, McDougal and Laird had all been arrested and were examined several times before the Sheriff. There was more than enough discrepancy for suspicion. Burke even blamed a stranger unnoticed by any of the others. But despite this and some circumstantial evidence, proof remained

Left: *Mary Docherty, also known as Mrs Campbell*

The report of Mary Docherty's dead body triggered the discovery of the West Port murders. The artist only had a brief description to work from, but her clothes, which were displayed at the trial, are likely to be accurately depicted.

Royal College of Physicians, Edinburgh. Image © National Museums Scotland.

Right: *Sir Robert Christison* by Dr John Adamson, calotype, mid-19th century

Robert Christison was Professor of Medical Jurisprudence (forensics) at the University of Edinburgh. He was one of several medical professionals who examined Mary Docherty's body but could not find conclusive proof she had been murdered and not choked accidentally.

thin. Elizabeth Main, Hare's servant, had seen her master carry a drunk elderly woman into a cellar. Mrs Ostler, one of the washerwomen who used Mrs Law's mangle, had disappeared, last seen entering Burke's building. A month after Mary Docherty's body was discovered, the Lord Advocate, responsible for the crown prosecution, was getting increasingly worried. He knew he had to take action soon and wrote the following a few weeks later, referring to himself in the third person:

> *After repeated and most anxious consideration of this extraordinary case, it appeared to the respondent that the evidence including the examination of Medical Gentlemen, was defective, both as to the fact of Docherty having been murdered, and as to who was the perpetrator of the deed. Conceiving it of the greatest importance, for the satisfaction and security of the public, that a conviction should be ensured, the respondent did not feel justified in hazarding a trial, on evidence which appeared to him to be thus defective. He well knew, from long experience, how scrupulous a Scottish Jury uniformly is, in finding a verdict of guilty where a capital punishment is to follow; and he deemed it hopeless to look for a conviction, where the fact of a murder having been committed was not put beyond the possibility of question.*[11]

Although there was not enough evidence to prove that Mary Docherty had been murdered, or indeed who had murdered her, the Lord Advocate felt it important, and of public interest, that a conviction was secured in this case. However, he knew from past experience that the jury would only give a guilty verdict if the proof was beyond doubt:

> *… Another consideration of still greater importance rendered this course indispensable. Some circumstances about this time transpired, which led the respondent to dread, that at least one other case of a similar description had occurred. In such circumstances he felt it to be his imperative duty, not to rest satisfied without having the matter probed to the bottom; and that he should for the sake of the public interest, have it ascertained what crimes of this revolting description had really been committed – who were concerned in them – whether all the persons engaged in such transactions had been taken into custody, or if other gangs remained whose practices might continue to endanger human life. Compared with such knowledge even a conviction for the murder of Docherty appeared immaterial. But such information could not be obtained by bringing to trial all the four persons accused. A conviction might lead to their punishment, but it could not secure such a disclosure.*[12]

The Lord Advocate badly wanted a conviction and doubted one would be secured on the evidence the police and Sheriff had discover so far. But more than that, he wanted the truth. How many people had been murdered by these criminals? Of even more importance was the question of whether these murders were unique or if other gangs were also murdering people for their bodies. Had the arrests been enough to stop these murders? Securing a conviction for the murder of Mary Docherty was of less importance than finding out the truth. Bringing the case to trial as it stood would not reveal the full extent of the murders.

He felt he had no choice but to attempt to secure a confession from one of the four prisoners, offering one of them the opportunity to turn King's evidence and secure themselves immunity for prosecution for acts they voluntarily confessed to. The only question was which of the four to select?

Helen McDougal was the obvious choice – she appeared to have been present throughout, but not to have taken the lead at any time in the sale of the bodies, nor to have been physically involved in the murders. Furthermore, a married person could not give evidence against their spouse and she was not legally married to Burke (although her counsel later argued that she was, with a Scottish irregular marriage), so her evidence could be used against all three of the others. But 'McDougal positively refused to give any information'.[13] Burke later said that Laird had suggested McDougal ought to be 'got out of the way', meaning murdered, and sold to Dr Knox because, being Scottish rather than Irish, she could not be trusted to keep their secrets.[14] However, on the matter of the murder she proved highly capable of keeping secrets. Margaret Laird had also refused and Hare could not be convicted on her evidence as she could not in law testify against her spouse. The Lord Advocate moved on to his final choice for informer – William Burke or William Hare.

Hare was eventually selected because, on the balance of the existing evidence, Burke appeared to be the principal villain. It was in Burke's home that the crime had taken place, and he had been the one to invite Mary Docherty to stay. Unlike the women, Hare took the offer to turn King's evidence and to make a statement about the murder of Mary Docherty, and any similar crimes in which he and Burke had been involved. Hare's statement does not survive, but reports state that it enabled the authorities to find other witnesses and to build cases against Burke for the murders of Mary Docherty, Jamie Wilson and Mary Paterson, and against McDougal for the murder of Mary Docherty only. Even with Hare's co-operation, the other murders had too little evidence to be taken to court. Margaret Laird was not tried, as Hare likewise could not give evidence against his spouse.

Jamie Wilson

Jamie Wilson was the second victim whose murder Burke, McDougal, Hare and Laird were charged with. He was killed about a month before Docherty and featured prominently in early speculation in the press about the number of murders that may have occurred. Jamie Wilson, widely known as Daft Jamie, was a familiar Edinburgh street figure who was now the subject of public concern. Had he been missing for weeks, or had he been seen after the arrests? Was it possible that such a well-known individual could have been dissected without being recognised? Of course all bodies dissected in anatomy rooms came from people who had been alive not long before, and many were those of people who had lived in or around Edinburgh. Recognition might not have been unprecedented or a reason not to dissect that body.

Much of the focus on Wilson's story arose because, despite the legal complications of Hare turning King's witness and being granted immunity from prosecution in return for evidence, there still remained the possibility of him being convicted of Wilson's murder or sued for his death. Despite legal examination of all potential avenues, and whether or not a private prosecution on behalf of Jamie's family could proceed, in the end it all came to nothing. Nonetheless Jamie, of all those murdered, received the most written attention. As a member of the respectable, unfortunate, poor, his death garnered more public sympathy than those deemed feckless or a drunkard, responsible for their own poverty.

Mary Paterson

Mary Paterson was murdered in April, although Burke did not recollect the date. She was described in contemporary newspapers as a 'well-known prostitute', the reports managing to combine horror at her murder with a degree of sanctimoniousness over the dangers of immorality and drink. Her friend, Janet Brown, described Paterson as 'irregular in her habits, but not so low as has been represented', and recent research has done much to remove her unjustified reputation.

Early that April morning, Brown and Paterson left the Canongate watch house, where they were probably sleeping off the effects of a night's drinking. They were imbibing a liquid breakfast when Burke encountered them. He invited the women to his lodgings. Rather than the West Port, they went to Burke's brother's house at Gibb's Close in the Canongate, where they had breakfast and a lot more whisky. Janet left the house for about twenty minutes and returned to find that Mary, whom she had left passed out on the bed, was no longer there. McDougal, Hare and Laird said that Paterson and Burke had gone out together. Paterson was never seen by her friend again.

Janet continued to enquire after Mary Paterson, but found no trace until, nearly seven months later, the newspapers announced the discovery of Mary Docherty's body, linked to the name of William Burke. She went to the police and was able to identify some of Mary Paterson's clothing among items found in Burke's house.

What had happened to Mary Paterson? Enquiries revealed that her body had been carried to Surgeons' Square only a few hours after her murder. One of the students, a 'tall lad', had recognised her. It was previously assumed that he knew her as a sex worker, but a woman known as Mary Paterson had recently been treated in Edinburgh's Royal Infirmary and it is possible he encountered her at the hospital. When the 'tall lad' asked Burke and Hare where they got the body, the two claimed they 'had purchased it from an old woman at the back of the Canongate',[15] supporting their story that they bought bodies.

Burke reported in one of his confessions that Mary Paterson's body was not dissected immediately, despite having been brought to Knox's rooms only a few hours after death. Instead, it was stored for months in whisky. This delay at first appears curious – the anatomists would surely have prized such a fresh body. And preservation in alcohol was far from perfect. It was used, for example, when bringing animal specimens from abroad, although they never arrived in perfect condition. However, for practical reasons more anatomical teaching, and thus dissection, took place in the winter when temperatures were naturally cooler, the dissection rooms less unpleasant and putrefaction slower.

Fresh bodies were so valuable that anatomists in London often employed people to get executed people's bodies onto the dissecting table as fast as possible after the execution – in some cases so quickly that reflexes were documented during the dissection process. Perhaps the delay over Mary Paterson was an indication that Dr Knox was in most need of a

Jamie Wilson, murdered October 1828

James Wilson was a well-known character on the streets of Edinburgh. Burke and the Hares accused each other of inviting Jamie in to drink with them before they murdered him.

Royal College of Physicians, Edinburgh. Image © National Museums Scotland.

Mary Paterson, murdered April 1828

William Burke invited Mary Paterson and Janet Brown for breakfast and whisky at his brother's house. Janet left while Mary was drunk and unconscious. She searched for her missing friend, but never saw her again.

Royal College of Physicians, Edinburgh. Image © National Museums Scotland.

supply for his teaching, and it was worth not making use of a body in excellent condition in the off-season in order to preserve it for the main dissection time in winter. Also highly significant is the question of how Burke knew that Mary Paterson was not dissected right away. Was there some degree of conversation with Dr Knox's men, suggesting that the relationship was beyond a brief financial transaction? Or had Burke and Hare simply been told that dissection would be delayed as justification for being given a lower, summer, rate of £8 for this particular body?

Thanks to Janet Brown's persistence in searching for her friend and going immediately to the police when the discovery of Mary Docherty's body became news, Mary Paterson's disappearance was one of the murders that most was known about.

On 4 December, William Burke and Helen McDougal were formally charged with the murder of Mary Paterson. By this time William Hare had confessed under King's evidence and neither Hare nor Margaret Laird were charged with the murder of Mary Paterson.

Left and right: William Hare and Margaret Laird

Margaret Laird and her baby were drawn by one of the jury at the trial, possibly Robert Jeffrey, whose profession was listed as engraver.

Royal College of Physicians, Edinburgh. Image © National Museums Scotland.

The trial

The trial of William Burke and Helen McDougal took place on Christmas Eve 1828 and lasted for 24 hours, through to Christmas Day – starting with several hours of legal argument about the charges. The trial attracted enormous legal interest, with eight lawyers volunteering to act for the defence in order to be part of such a high-profile case. Burke was charged with all three murders, but McDougal only with that of Mary Docherty, raising the question of whether two people could be tried in the same trial on such different charges? The four judges decided not, and only the charge of the murder of Mary Docherty was heard, to which both Burke and McDougal pleaded not guilty. If Burke were not to be convicted for Docherty, the option remained to try him again for the murders of Jamie Wilson and Mary Paterson.

From a modern perspective, a startling omission from the trial was Mary Docherty's son, Michael Campbell. Docherty had come from Glasgow to look for her son, and arrived at his lodging place on 30 October. It was said that

Left and right: William Burke and his partner Helen McDougal

Royal College of Physicians, Edinburgh.
Image © National Museums Scotland.

he left that day – there is probably an incorrect 'Monday' here in the trial transcript. However, as Campbell paid for Mary's lodging that night, the implication is that she had found him, but was rejected by him. She left the next morning saying to his fellow lodger that she was leaving town and did not know where her son was. He was not called as a witness, nor interviewed in the press. Instead, evidence about her identity was submitted at the trial by Campbell's landlady and a fellow lodger, Charles McLauchlan, who had known both mother and son in Donegal.

William Burke was convicted of the murder of Mary Docherty on the existing evidence. The verdict arrived at half past eight on Christmas morning, after the Scottish jury of 15 men had deliberated for fifty minutes, although two jurors apparently had doubts about his guilt. Helen McDougal was given the benefit of the doubt. With the particularly Scottish verdict of 'not proven', the jury did not find her innocent but rather the crown had failed to prove her guilt. This was the verdict the Lord Advocate had feared for all four prisoners, without a confession from one of them.

McDougal was released and attempted to go home to the West Port area, whereupon she was recognised and the police summoned to take her back into protective custody. William Hare and Margaret Laird, however, were returned to prison. As a self-confessed mass murderer, the immunity Hare had been promised was about to be tested to its utmost and all possible avenues for his punishment explored.

Burke alone was sentenced to be executed, 'to be hanged by the neck by the hands of the common executioner upon a gibbet until he be dead and his body thereafter to be delivered to Dr Alexander Monro Professor of Anatomy in the University of Edinburgh to be by him publicly dissected and anatomized'[16] on Wednesday 28 January. The Lord Justice Clerk David Boyle presiding made an unusual addition to his remarks before the formal sentencing, stating that he hoped 'if it is ever customary to preserve skeletons, yours will be preserved in order that posterity may keep in remembrance your atrocious crimes'.[17]

Confessions

After William Burke was found guilty, he confessed to the murder of Mary Docherty and to 15 more – or rather to 14 more, mentioning one instance in which Hare had acted alone. These confessions were mostly taken down from his words, but some portions were written by Burke himself as he was, unusually for the time, literate. It is from these confessions that most of what we know about Burke, and about the crimes, comes. They were compared with Hare's statement and found to agree on the main points, which was considered a strong argument for the truth of their content.

Opposite: Robes belonging to David Boyle, Lord Justice Clerk

David Boyle, Lord Justice Clerk, presided over the four judges at the trial of William Burke and Helen McDougal. As Scotland's senior criminal judge, his presence was a clear sign of how significant the trial was considered to be. These are the silk robes of the Court of Justice that David Boyle would have worn during the trial in the criminal court.

On loan from Kelburn Castle.
Image © National Museums Scotland.

Above and opposite: Handwritten confession by William Burke, 27 January 1829

While awaiting execution in prison William Burke dictated confessions giving details of the murders, as far as he remembered them. Burke was literate and this final confession, written on the day before his execution, appears to be in response to some final or much-repeated questions.

> Dier friend i mean to mention to you some things that i was acused for first conserning a snuff box that was supposed to be daft Jemmys i take it to Death with me that the snuf box that i had in the Lockup and was after convied to the Sherifs Office [was] Not James box – i was my own, the spoon was Jameys Wm Burk.
>
> Next it was reported that i had Doctors instements in my house and that there was a Letter found in my house simmingly from a Doctor …… i Also Declare it false and there was questens put to me Concerning murders in ……

I gave the Sheref an acount of Sixteen persons who had been murdered by hear [Hare] and me Which was all i ever had any hand in all my life –

William Burk –

Conserning Daft Jemy i Declare that Wm hear was the first that laid hands on him i never thretined to stab him with a knife ……

Both images: Courtesy of the New York Academy of Medicine Library

Life masks of William Burke and William Hare

The pseudoscience of phrenology was at its height of popularity at this time. Phrenology attempted to correlate personality with the shape of a person's skull, and of the brain within the skull. Because of this, the shapes of Burke and Hare's skulls were of great interest in trying to answer the question about what makes a murderer. These life masks were taken in prison.

© Anatomical Museum, The University of Edinburgh

… the information which Hare gave to the sheriff on the 1st December last, (while he imputed to Burke the active part in those deeds which the latter now assigns to Hare,) Hare disclosed nearly the same crimes in point of number, of time, and of the description of persons murdered which Burke has thus confessed; and in the few particulars in which they differed, no collateral evidence could be obtained calculated to show which of them was in the right.[18]

It is not surprising that William Hare's confession was shaded to implicate Burke, and Burke's to implicate Hare and spare Helen McDougal blame. Margaret Laird is also presented as highly complicit in Burke's narrative as she insisted on one pound from the sale of each body as her share for the use of her house, and it was she who lured Jamie Wilson to her lodging house and to his death.

Burke's confessions further present a fairly clear picture of the sale of the body of Donald who died in the lodging house, and the financial temptation this presented. However, there is a lot of information missing, especially about the victims – suggesting that to the murderers they were merely a commodity and not remembered as people. There is doubt especially about the order of the murders. Hare's account was not the same as the two slightly differing orders Burke provided – and Dr Knox offered no help whatsoever to confirm which account was correct.

The unprosecuted murders

Little is known about the other people murdered by Burke and Hare. The only information is derived from Burke's confessions, although he was not always able to identify them. If their full names were discovered – either through the formal investigation or as recognised by friends – they were not fully reported. When these murders were investigated, emphasis was placed on the conviction of the malefactors rather than justice for each individual victim, so only the most promising routes were followed and investigation of unnamed victims was perhaps limited. Or perhaps some had simply vanished out of a mobile population without their passing being noted. It is also likely that the horror of the financial motivations behind the murder was of more interest to the city authorities, and to a literate, newspaper-buying population, whereas the details of the lives and deaths of poor people was not. In other words, the *selling* of the bodies was of more interest than the murders.

However, it has been established that the early victims were either ill or were rendered insensible through drink. Joseph, for example, was a miller who fell ill of a fever in the Hares' lodging house. News of sickness in the dwelling was bad for business, which increased the temptation of Burke and the two Hares to murder him and sell his body. Abigail Simpson had come in from Gilmerton, then a village outside Edinburgh, to hawk salt on the street. She lodged at the Hares and drank with her hosts. When she fell unconscious, she was easy to smother. The bodies of Joseph and Abigail were carried to the dissecting rooms by porters.

Burke was inconsistent in his account of which of these people was the first to be murdered, triggering the idea that lodgers, or others, could be similarly killed and sold.

The next victims were also vulnerable and weak. They included a forty-year-old English matchseller from Cheshire who had lodged at the Hares' for a few nights. He was apparently ill with jaundice. Burke never discovered his name, nor that of a very drunk elderly woman who had been persuaded by Margaret Laird to stay with them for the night.

As their experience grew, Burke and both Hares plied more lodgers with drink and grew less cautious. Effy was a cinder gatherer, one of many who made a living from other people's waste. She had sold scraps of leather to Burke for reuse. She had also lodged at the Hares' house.

One evening, Burke spotted some policemen dealing with a drunk woman. He offered to see her home and they gladly handed her over. She was never seen again. Another lodger, Mary Haldane, was plied with drink and later smothered. Some months later, her daughter Margaret became a victim. Burke also mentions another lodger, murdered by Hare while Burke was away.

About midsummer, a woman and her disabled grandson of about twelve came to lodge and were both murdered. Their bodies were crammed into a barrel and taken to the dissecting rooms in Hare's cart. The horse stopped at Meal Market on the Cowgate and refused to go any further, forcing Burke and Hare to employ a porter with a barrow for the last part of the journey. The pair were somewhat spooked by the horse's behaviour. The animal was a recent acquisition and Hare decided it was run down and useless and had it shot by the local tanners for the value of its hide. Hare attempted to make the hide appear more valuable by disguising two wounds or scars on the horse's shoulder. As Burke said:

> *Hare was taken in by the horse he bought that refused drawing the corpse to Surgeon Square and they shot it in the tan-yard. He had two large holes in his shoulder stuffed with cotton and covered over with a piece of another horse's skin to prevent them being discovered.*[19]

Mrs Ostler was a washerwoman who hired the use of a mangle owned by Burke's neighbour, Mrs Law, in the building he moved to after moving out of the Hare's property. She too fell victim to murder. And even Ann Dougal, a cousin of Helen McDougal, was not spared the fate of all the others.

Burke's execution and dissection

William Burke was hanged in the Lawnmarket at quarter past eight on the morning of 28 January 1829, just over a month after his trial. Every window overlooking the execution site had been hired and, despite the pouring rain – not obvious in the engravings of the scene – the street was packed from the early hours of the day.

It is estimated that a crowd of 20–30,000 people assembled to see Burke executed. The crowd, it was reported, waited mostly in solemn, and likely sodden, silence until the prison officers, Catholic priests and Burke appeared. During the execution there was a clamour among the onlookers for Hare and Knox to be hanged too.

Burke had complained that Robert Knox had not paid in full for Mary Paterson's body and if he were to be paid the remaining £5 he would be able to afford a smarter coat for his public appearance. Before his execution, Burke was reported to have seemed resigned and prepared for his death. The accounts of his behaviour do not include him making any comment on the dissection of his body, which was to follow as part of his sentence.

The execution of William Burke, 1829, by Walter Geikie

Prints of William Burke's execution were rapidly produced and sold to the fascinated public. It is possible that they were even drawn beforehand in preparation and anticipation of the scene.

National Galleries of Scotland

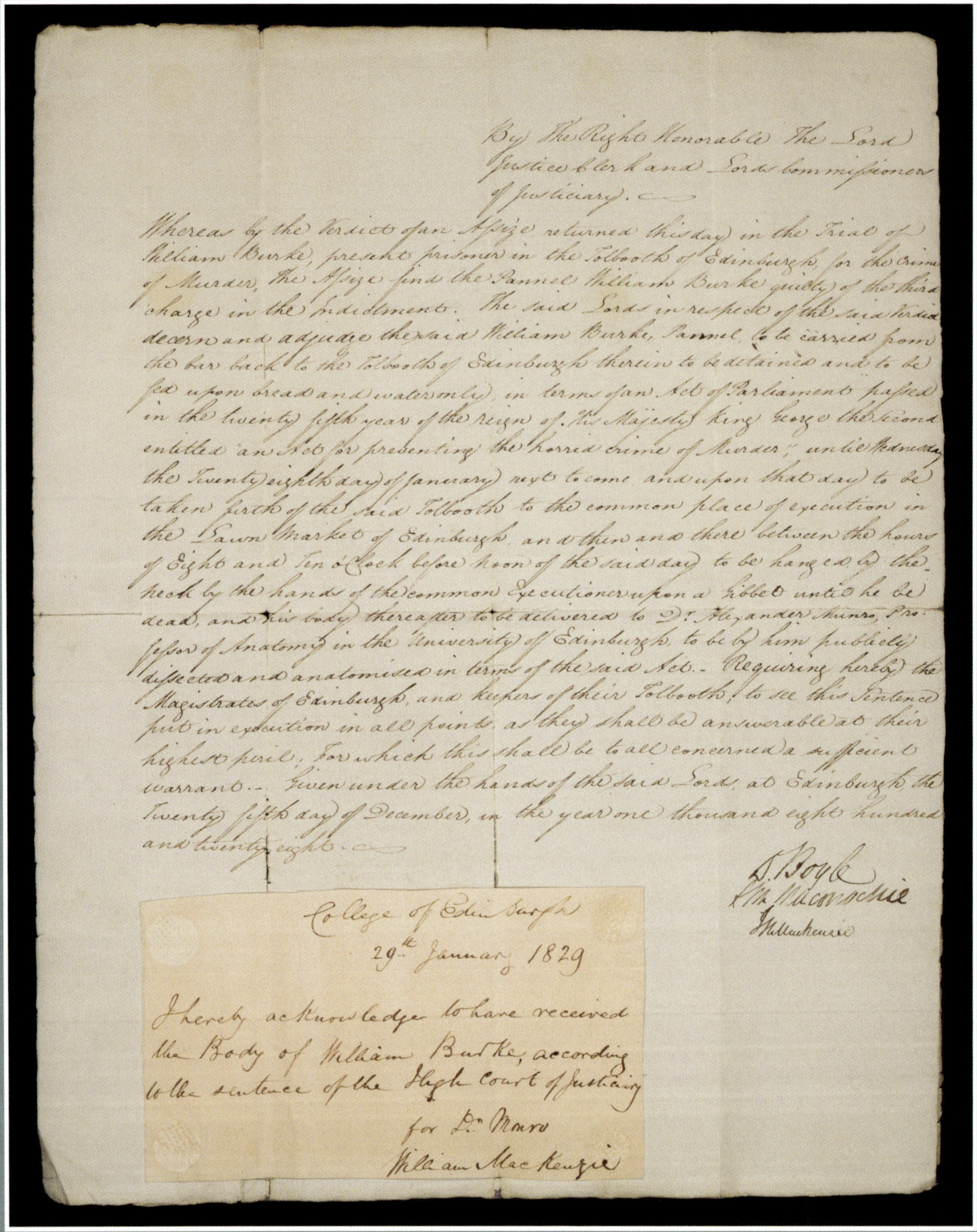

By the Right Honorable The Lord Justice Clerk and Lords Commissioners of Justiciary.

Whereas by the Verdict of an Assize returned this day in the Trial of William Burke, present prisoner in the Tolbooth of Edinburgh, for the crime of Murder, the Assize find the Pannel William Burke guilty of the third charge in the indictment. The said Lords in respect of the said Verdict decern and adjudge the said William Burke, Pannel, to be carried from the bar back to the Tolbooth of Edinburgh therein to be detained and to be fed upon bread and water only, in terms of an Act of Parliament passed in the twenty fifth year of the reign of His Majesty King George the second entitled an Act for preventing the horrid crime of Murder, until Wednesday the Twenty eighth day of January next to come, and upon that day to be taken forth of the said Tolbooth to the common place of execution in the Lawn Market of Edinburgh, and there and there between the hours of Eight and Ten o'Clock before noon of the said day to be hanged by the neck by the hands of the common Executioner upon a Gibbet until he be dead, and his body thereafter to be delivered to Dr Alexander Munro, Professor of Anatomy in the University of Edinburgh, to be by him publicly dissected and anatomised in terms of the said Act. Requiring hereby the Magistrates of Edinburgh, and keepers of their Tolbooth, to see this Sentence put in execution in all points, as they shall be answerable at their highest peril, for which this shall be to all concerned a sufficient warrant. Given under the hands of the said Lords at Edinburgh the Twenty fifth day of December, in the year one thousand eight hundred and twenty eight.

D. Boyle
Jn. Macconochie
H. Mackenzie

College of Edinburgh
29th January 1829

I hereby acknowledge to have received the Body of William Burke, according to the sentence of the High Court of Justiciary for Dr Monro

William MacKenzie

Death warrant for William Burke and receipt for the body dated 29 January 1829

The death warrant for William Burke directed that his body was to be delivered to Alexander Monro to be 'publicly dissected and anatomised'. A receipt for his body from Monro's assistant, dated the day after his execution, is fastened to the death warrant.

On loan from Edinburgh City Archives. Image © National Museums Scotland.

The day after William Burke was hanged, his body was taken to the anatomy room of the University of Edinburgh. There Alexander Monro *tertius* dissected and lectured upon his brain, an unusually bloody dissection even for an anatomy theatre. A few days later, Monro used some of the blood to write a note. While he was accustomed to dissecting the bodies of people hanged for murder, this case was different, even for the anatomist. To calm unrest from the many students and townspeople unable to attend the dissection, Burke's body was laid on the dissection table while students filed past that afternoon. Thousands of the public were admitted the next day.

Not everyone was keen to be part of this ghoulish spectacle. Sir Walter Scott wrote in his diary:

The corpse of the Murderer Burke is now lying in state at the College, in the anatomical class, and all the world flock to see him. Who is he that says that we are not ill to please in our objects of curiosity? The strange means by which the wretch made money are scarce more disgusting than the eager curiosity with which the public have licked up all carrion details of this business.[20]

Burke's head and brain were of particular interest in the bid to understand what makes a murderer. Was there anything dissection could find in the brain, or about the shape of the brain or the skull around it? The pseudoscience of phrenology, which sought to relate personality and ability to the shape of someone's skull, was prominent in Edinburgh and casts were taken of both Burke and Hare's heads for this (page 82), as well as a cast of the inside of Burke's skull, showing the shape of his brain.

In accordance with the suggestion of the Lord Justice Clerk, David Boyle, William Burke's skeleton was cleaned and mounted and preserved in the anatomical museum of the University of Edinburgh. It remains on display today.

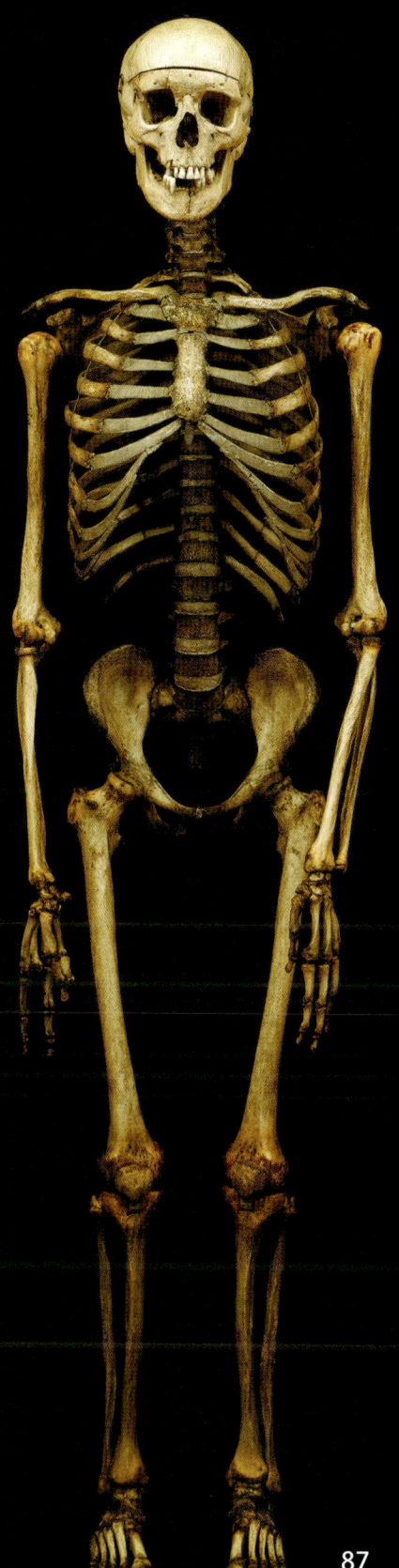

Skeleton of William Burke, 1829

Burke's skeleton was cleaned and assembled, at the suggestion of David Boyle, the Lord Justice Clerk, as part of his sentence. Boyle had stated, 'And I trust that if it is ever customary to preserve skeletons, yours will be preserved in order that posterity may keep in remembrance your atrocious crimes'.

© Anatomical Museum, The University of Edinburgh

McDougal, Hare and Laird

Helen McDougal was released after the trial and attempted to return home to the West Port not realising how impossible this would be. She was hounded out and left Edinburgh. Margaret Laird went to Glasgow with her baby. She too needed to be rescued from a mob and was escorted onto a ship to Ireland to get her out of Scotland to preserve the peace.

William Hare, when the authorities reluctantly concluded there was no way to convict or punish him for his crimes, was put on a coach going south. In Dumfries, another passenger announced who he was and he had to be smuggled out of town for his own protection, leaving on foot without any belongings. This was the last reliable sighting of him. There were contradictory reports of several different horrible fates for him – or that he had quietly managed to reach Ireland.

Riot at Dumfries! Hare's Arrival, newspaper cutting, 1829

William Hare had to be released after no legal route could be found to bring him to justice. He was put on a coach out of Edinburgh, but was recognised in Dumfries and smuggled out of the town.

Royal College of Physicians, Edinburgh. Image © National Museums Scotland.

Robert Knox

How significant were the West Port murders for the anatomists? As a supply of bodies, they were only one of several suppliers and only for a single year. While it is not known exactly how many bodies Dr Knox purchased in 1828, records from subsequent years reveal that these 16 murdered people were probably about a quarter of the bodies purchased by his anatomy school in that year. Although this additional source of bodies may have been a welcome addition to Knox's supply, it was not enough to change significantly the way his research or school operated, and had no effect on the many other anatomy schools in Edinburgh. The vast majority of bodies traded to the dissection tables still came from grave robbing or appropriation as described in the previous chapter.

In 1828 the murders of subjects for anatomical dissection did not revolutionise the practise of anatomy. In 1829, however, public knowledge of these murders was extremely significant for Knox and had a lasting impact on his career. Despite an inquiry of his peers concluding that he could be considered innocent of wrongdoing, there was still a strong desire of institutions to distance themselves from him, with the public still of the opinion that he should have been held more to account.

Notably missing from the investigation were accurate records from Knox's business of the bodies he had bought. Had he been so disinterested in the people whose bodies formed the basis of his practice that no records were kept? Was this intentional to reduce the possibility that irate relatives would identify and reclaim grave-robbed bodies? It appears so, for much of the details we know about Knox's business came from records established by his assistants from 1829. This was presumably in response to the censure that came about from Knox's inability, or perhaps complete refusal, to provide detailed information about all the bodies purchased from Burke and Hare and the exact date of each transaction.

Robert Knox was neither arrested nor charged. Although summoned to attend Burke and McDougal's trial as a potential witness, he was one of several who remained uncalled. Perhaps his evidence was only thought significant for one of the other charges, the murders of Jamie Wilson and Mary Paterson, that were not in the end tried. He did, however, face considerable blame in the public press – surely an anatomist of his experience could recognise murdered bodies?

A panel of Knox's peers was assembled to enquire informally into his conduct and they exonerated him. As the medical men testifying at Burke's trial had been unable to prove conclusively that Mary Docherty had been murdered, the other bodies could not be assumed to have shown plainer signs. Knox maintained that he believed his suppliers were buying bodies of people who had died without anyone who cared enough to see them safely buried – exactly the situation of the first body he had purchased from Burke

Dr Robert Knox (1791–1862), by David Octavius Hill and Robert Adamson, calotype, 1840s

Despite Robert Knox buying the murdered bodies, an enquiry of his peers found him innocent of wrongdoing. However, his reputation in Edinburgh never recovered. He was pushed to resign from his responsibility for the collections of the Royal College of Surgeons; and while he initially remained a popular anatomy teacher, the University made their anatomy course compulsory for students, which ruined his business.

National Galleries of Scotland

and Hare, the pensioner Donald who died of natural causes in the Tanner's Close lodging house.

Sir Walter Scott was invited to form part of the peer enquiry but refused. His own private writings make it clear that he held Knox culpable. Scott had trained as a lawyer and wrote his thesis for admission as an advocate on '*De cadaveribus damnatorum*' [the bodies of the condemned].

In the short term, however, Dr Knox maintained his popularity as a teacher. It did not matter that the people of Edinburgh hanged an effigy outside his front door and smashed windows; or that the press and satirists vilified him. His students remained his advocates and protectors, still signing up to his classes in large numbers. The medical establishments, however, despite his exoneration, did not continue to welcome him. The Royal College of Physicians considered 'Burke's affairs' and sought to distance themselves from the murders in the public mind. The Royal College of Surgeons of Edinburgh pressurised Knox into resigning from his post in charge of their museum and the collection of specimens he had largely established for them – which he finally did in 1831. The University professors flatly refused to have him appointed as one of their number, threatening to resign en masse should his application to become Professor of Pathology be even considered. Further, in the 1830s the University made its own practical anatomy class compulsory for students, meaning that any external anatomy course a student attended would need to be in addition to the University's course rather than as an alternative choice. This finally reduced the numbers signing up to Knox's classes, curtailing the bulk of his income.

Dr Robert Knox finally left Edinburgh after the death of his wife. He failed to establish himself as a teacher in Glasgow and then London.

Notes

1. Chadwick 1842, p. 411.
2. Buchanan 1829, p. 43.
3. '*Noctes Ambrosianae*' 1829, in *Blackwood's Edinburgh Magazine*, p. 401.
4. Buchanan 1829, p. 32.
5. *Ibid.*, p. 44.
6. Ireland 1829, pp 42–43.
7. *Ibid.*, p. 76.
8. *Ibid.*, p. 75.
9. Buchanan 1829, p. 85.
10. *Ibid.*, p. 30.
11. *Ibid.*, p. 11.
12. *Ibid.*, pp 11–12.
13. *Ibid.*, p. 12.
14. *Ibid.*, p. 42.
15. *Ibid.*, p. 35.
16. *Ibid.*, p. 199.
17. *Ibid.*
18. *Ibid.*, p. 32.
19. *Ibid.*, p. 39.
20. Scott 1890/91, journal entry for 31 January 1829, p. 227.

Opposite: Miniature coffins and figures, 1830s

In June 1836, a group of 17 miniature coffins were found hidden on Arthur's Seat in Edinburgh. One explanation for these mysterious coffins and figures is that they commemorated the 16 people murdered by Burke and Hare, and 'Donald' whose body was the first to be sold.

An end to grave robbing

The discovery of the West Port murders changed neither the issues nor means of body supply. In 1832, George Wilson, later the founding Director of the Industrial Museum of Scotland, one of the museums that developed into the National Museum of Scotland, began medical studies at Edinburgh University at only 14. He wrote about an early recollection 'of a fellow-student lying under charge of the police in a surgical ward of the Infirmary, with a gunshot wound received in a resurrectionary expedition to Musselburgh churchyard. He was looked upon as a martyr by the students, and as little better than a murderer by the people. The law dealt mildly with such transgressors, and, til roused by the hideous murders in Edinburgh, was, in regard to resurrectionism generally, Justice without the scales, and with a very thick bandage over her eyes'.[1] Evidently grave robbing still continued, carried out personally by students as well as those undertaking it for financial gain. The other source of bodies was executed people, and Wilson states that 'the more thoughtless among the students anticipated a public execution with a certain grim satisfaction and professional interest'.[2]

It had been acknowledged for years that change was overdue: anatomists appropriating, or buying, bodies was a bad method of supply for all involved. It did not provide anatomists with good material, it caused widespread upset among those whose bodies might be taken, and it channelled money to grave robbers. In March 1828, a petition of 248 signatures was presented by Edinburgh University students to the Commissioners inspecting the Universities and Colleges of Scotland. It requested legislation to provide more bodies for dissection, explaining that their scarcity impeded the study of medicine and caused many to choose to study instead in Paris or other foreign universities that were better supplied.

A parliamentary enquiry was also taking place into the provisions of bodies for dissection. Legislation was placed before parliament in 1828, but it failed to pass.

In 1831, John Bishop, a grave robber who estimated he had sold 500–1000 bodies to anatomists, and James May, sold the fresh body of a 14-year-old boy to King's College School of Anatomy in London. Suspecting the body had not been buried, and perhaps with suspicion aroused by the West Port murders, Herbert Mayo, the Professor of Anatomy, summoned the police. This uncovered what became known as the London Burkers. After their conviction, Bishop, May and a third man, Thomas Williams, confessed to killing three people in order to sell their bodies. Mayo, the anatomist who had reported his suspicions, took the opportunity to push for parliament to act to prevent future murders.

In 1832, an Act for Regulating Schools of Anatomy was passed, commonly known as the Anatomy Act. It enabled anyone who had legal possession of a body to pass it to anatomists for dissection, provided no relative objected. The bodies of people executed for murder were no longer to be dissected, but people who died in hospitals, poor houses, asylums, prisons and other institutions were likely to be. It was argued that this changed dissection from a punishment for murder to a punishment for being poor. Politician William Cobbett was among those who objected: 'They tell us it was necessary for science. Science? Why, who is science for? Not for poor people. Then if it is necessary for science, let them have the bodies of the rich, for whose benefit science is cultivated.'[3]

The Anatomy Act only made bodies available to anatomists who were licenced according to the Act, and established that the anatomy schools were responsible for record-keeping and burying the bodies after dissection, and all costs associated with their supply. The Act did not immediately operate smoothly. It did not provide as many bodies as the anatomists hoped for, or prevent abuse. While relatives could in theory object to bodies being sent for dissection, their consent was not required, and they might not be informed of the death in time. Some institutions held relatives who withheld assent responsible for the cost of the burial and refused to acknowledge objections from people who could not afford it. The Act, however, did finally cut off Dr Robert Knox's supply of bodies. The licensed anatomists at the University of Edinburgh refused to give him access to any.

The Act required that bodies were buried after dissection. Knox's brother Frederick complained about this in a book published in 1836 when describing how to prepare and preserve a human skeleton for study or display: 'At the moment it is actually contrary to the express letter of the law to make any such preparation.'[4]

The Act also allowed for people to donate the bodies of their relatives, or their own bodies as long as relatives did not object. Initially very few people did this. Most bodies used for dissection throughout the rest of the nineteenth and into the twentieth century were unclaimed ones. At first many came from people who died in poor houses, but as medical and social care changed so did the locations where people lived and died, and a greater proportion of bodies used for dissection came from people who had died in asylums.

Over the twentieth century, attitudes to dissection gradually changed and universities began to encourage people to donate their bodies. Cremation became more common, and blood and organ donation raised the profile of donation. An increase in body donations followed the formation of the National Health Service, with some people seeing this as a way to give back to an organisation that had supported them in life. Today, all the bodies used by anatomists in the United Kingdom are donated, given with consent; and the legislation no longer permits unclaimed bodies to be taken and used.

How anatomy is studied has changed considerably since the sixteenth century, but the human body remains at the centre of this study and new anatomical discoveries are still being made. Understanding the human body is even more important now as medicine has more capabilities to make use of that understanding for practical treatments or diagnosis. The dissection of human bodies is still a vital part of learning anatomy, medicine and surgery. Those who choose to donate their bodies are acknowledged with gratitude and remembered as silent teachers.

Notes

1. Wilson 1861, p. 92.
2. *Ibid.*
3. Crowther 1965, p. 9.
4. Knox, F. 1836, p. 41.

Further reading

Archives and Manuscripts Collections, University of Edinburgh, EUA/IN1/ACU/A2/21/1.

Audubon, Maria R. and Elliott Coues 1899. *Audubon and his Journals: with Zoological and other Notes by Elliott Coues; in 2 Volumes* (New York: Charles Scribners).

Bell, Charles, Surgeons' Hall, The Royal College of Surgeons of Edinburgh [catalogue entry].

Blancken, Gerrard 1704 (revised). *A catalogue of all the cheifest rarities in the publick theatre and Anatomie-hall of the University of Leyden* (Leyden: Hubert vander Boxe, 1697).

Buchanan, Robert 1829. *Trial of William Burke and Helen M'Dougal before the High Court of Justiciary …* (Edinburgh: Robert Buchanan).

Caledonian Mercury, 4 May 1741, 31 May 1742.

Chadwick, Edwin 1842. *Report on the sanitary condition of the labouring population of Great Britain* (Edinburgh: Edinburgh University Press, 1965).

Clayton, Martin and Ron Philo 2017. *Leonardo da Vinci: Anatomist* (London: Royal Collection Trust; first edition 2012).

Crowther, J. G. 1965. *Statesmen of Science* (London: Cresset Press).

Darwin, Charles (ed. Francis Darwin) 1892. *His Life told in an Autobiographical Chapter, and in a Selected Series of his Published Letters* (London: John Murray).

Dingwall, Helen 1995. *Physicians, Surgeons and Apothecaries: Medical Practice in Seventeenth-Century Edinburgh* (East Lothian: Tuckwell Press).

Edwards, Owen Dudley 2014. *Burke & Hare* (Edinburgh: Birlinn Ltd).

Ireland, Thomas 1829. *The West Port Murders; or an authentic account of the atrocious murders committed by Burke and his associates; containing a full account of all the extraordinary circumstances connected with them* (Edinburgh: Thomas Ireland Jnr).

Jones-Lewis, Molly A. 2018. 'The Case of the Missing Ban: Cadaver Dissection in Roman Law' (USA: The Classical Association of the Middle West and South) <https://camws.org/sites/default/files/meeting2018/abstracts/249.CadaverDissectionRomanLaw.pdf>

Knox, Frederick John 1836. *The Anatomist's Instructor, and Museum Companion* (Edinburgh: Adam and Charles Black).

Langenwagen, Maurits 1766. *Naam-lyst van alle persoonen, die binnen Amsterdam sedert het jaar 1693 tot 1766.*

Levine, Jeffrey M. 2014. 'Jewish History in Vesalius's Fabrica' <http://jmlevinemd.com/jewish-history-vesalius-fabrica/>

Maclise, J. 1841/44. *The circulatory system* (with coloured lithograph by J. Maclise) (London: Taylor & Walton).

Moore, Wendy 2006. *The Knife Man: Blood, Body Snatching, and the Birth of Modern Surgery* (Broadway Books).

Mossop, Dave [data]: <http://www.capitalpunishmentuk.org/>

'Noctes Ambrosianae' 1829, in *Blackwood's Edinburgh Magazine.*

Pepys, Samuel, 1668. 'The Diary of Samuel Pepys' (Friday, 3 July 1668). <https://www.pepysdiary.com/diary/1668/07/03/>

Postle, Martin 2004. 'Flayed for art: écorché figure in the English art academy', *British Journa*l 5, no. 1.

Richardson, Ruth 2001. *Death, Dissection and the Destitute.* (Chicago: University of Chicago Press).

Rosner, Lisa 2009. *The Anatomy Murders: Being the True and Spectacular History of Edinburgh's Notorious Burke and Hare …* (Pennsylvania: University of Pennsylvania Press).

Scott, Walter 1890/91. *The Journal of Sir Walter Scott: From the Original Manuscript at Abbotsford* (entry for 31 January 1829) (Edinburgh: David Douglas).

Vasari, Giorgio 1568 (1907). *Lives of the Most Excellent Painters, Sculptors, and Architects* (trans. Louisa S. Maclehose) (New York and London).

Vesalius, Andreas 1543. *De humani corporis fabrica* [*On the Fabric of the Human Body*].

Waterhouse, George Robert 1841. *The natural history of Marsupialia or pouched animals,* 'The Naturalist's Library', vol. 25 (Edinburgh: W. H. Lizars).

Wilson, George 1861. *Memoir of Edward Forbes, F.R.S. Late Regius Professor of Natural History in the University of Edinburgh.* (London: Macmillan).

Acknowledgements

An exhibition and publication is far more than the work of one single person. Much of this combined effort took place during the challenging times and reduced onsite working of Covid-19 restrictions. I am very grateful to the many institutions – universities, professional colleges, museums and collections – who have lent objects for this project, preparing them for display and generously sharing knowledge about their collections and making images available.

I am also very grateful to the many colleagues, both at and outwith National Museums Scotland, whose outstanding work has made this project what it is:

> the registrars who organised the practical aspects of borrowing objects; the conservators who prepared about 230 objects for display, and the photographers who imaged them; collections care colleagues who organised quarantine of incoming loans, packed the collections for safe transport, and ensured all objects were where they needed to be for each stage of exhibition preparation; the mount-makers and designers who brought the exhibition to fruition; those who bring it to public attention through press and marketing; the many colleagues who welcome our visitors to the exhibition and events; those who designed, edited and sourced the images for this publication, and to Sophie Goggins and Julie Gibb who read the chapter drafts; to all the colleagues from Scottish History and Archaeology, Science and Technology, and NMS Exhibitions and Design departments on the changing exhibition content team, who developed the stories and text for the exhibition on which this publication draws extensively; and, above all, to the Exhibition Officer, Emma Ulloa, who co-ordinated both the intellectual and practical content, ensuring quality, accuracy and efficiency across the whole project.

Tacye Phillipson
Senior Curator of Science
National Museums Scotland, Edinburgh 2022

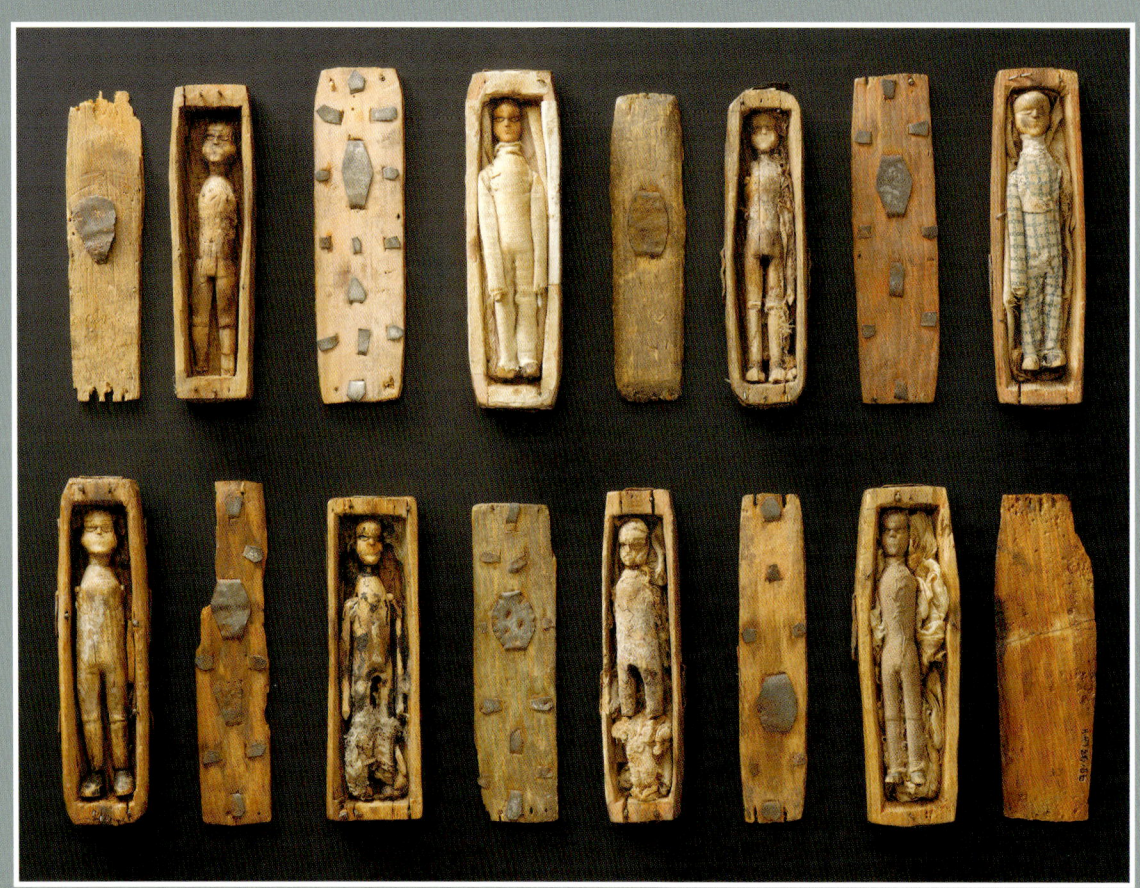

Miniature coffins and figures, 1830s